British
Ingenuousness

Juleon M. Schins

authorHOUSE®

AuthorHouse™ UK
1663 Liberty Drive
Bloomington, IN 47403 USA
www.authorhouse.co.uk
Phone: 0800.197.4150

Published by AuthorHouse   12/19/2015

ISBN: 978-1-5049-9694-5 (sc)
ISBN: 978-1-5049-9693-8 (hc)
ISBN: 978-1-5049-9695-2 (e)

Synopsis

One often hears: "Believers and agnostics alike, everybody seems convinced that God (or something out there) has but little influence on our material world: the more we know scientifically, the less room is left for God to influence material reality."

Well, the opposite is true.

The more we scientifically know about our material world, the more God seems to run everything down here, with us humans filling some tiny left-over gaps. The belief that everything is matter, and that only humans (and eventually animals) make choices with material consequences, is incompatible with quantum mechanics, but widely accepted among the ingenuous. Ingenuousness is roughly identical throughout the occidental world, be it Christian or ex-Christian. Yet every country has its own specific traits. That is the reason this book, directed to the British, is called *British ingenuousness*; however, the title is not at all meant to suggest that the British are more naïve than any other people.

The book contains two parts: five chapters and five appendices. The five chapters describe the author's view on the social consequences of occidental ingenuousness, and the appendices the philosophical roots of that ingenuousness.

The chapters describe typical aspects of occidental ingenuousness: that of young girls, who think their sexy looks enhances their odds to find love; that of clergymen, who seem to think we still live in the Garden of Eden; that of legislators, who do not manage to reduce British recidivism below 26%, with on average 3 offences per reoffender; that of adult women, who have not the slightest idea how to remedy gender discrimination; that of retrograde ecologists, who believe carbon dioxide is poisonous, that our universe was optimal in 1980, and should not change any more; and finally, though most demanding to grasp, the ingenuousness of adults, who believe we're not doing so bad after all.

Biography

The author was born 1964 in Sorengo (Switzerland) from Dutch parents. He was raised in Italy, obtained his high school degree 1982 from the "Scuola Europea di Varese", his master's 1987 in optics at the University of Amsterdam, his doctorate 1992 in molecular physics. He had some post-doc experience on atomic physics at the École Nationale de Techniques Avancées in Paris, and on biophysics at the University of Twente. From 2002 the author teaches nanotechnology at the Delft University of Technology, and is specialized in Optical and teraHertz spectroscopy of semiconductor nanocrystals.

The author has always been interested in the relation between physics and philosophy. He wrote books and articles claiming that quantum mechanics and causality can be understood consistently and univocally in both physics and philosophy by means of Aristotle's hylomorphism.

Contents

Introduction

Fourteenth Century, rural France. There is quite some confusion in the central square of the town, where an execution is being prepared. The family members shout out, "Arrêtez-le!" ("hold him"), to prevent the priest from reaching the condemned. The latter has eyes only for the priest, and pays hardly any attention to the executioner's axe. The bored executioner is visibly irritated because of the umpteenth delay, and sways his axe impatiently. When the condemned realizes the priest is too strongly withheld, he loses his theatrical dignity: stricken with panic, he curses the relatives of his victim, threatening them with divine justice.

Modern enlightened people poke fun at such medieval scenes. The Middle Ages had everything upside down: the condemned appeals to divine justice, while the relatives of the victim harass the clergyman.

The key to understanding the scene is the medieval faith in the existence of a spiritual world, a faith as solid as granite. Both the condemned and the victim's family believe the convicted will go straight to hell if deprived of the last sacraments. To medieval people the sacramental confession, divine justice, the extremes, the devil – were 'things' no less real than any other tree in the square. Since then a lot has changed. The Church has changed, in that it admits baptism of desire, perfect contrition, and a much less severe praxis of penance. Society has changed, too, as it went through various revolutions: the industrial revolution of the 18th century, the technological revolution of the 19th century, the global-economic revolution of the 20th century, and the information revolution of the 21st century. All those revolutions brought us excellent things, like hygiene, cheap food, and beautiful houses to live in. There is but one thing that we've gradually been losing: the granite conviction of the existence of a spiritual world. Dreadful sorry to say so, but the silly ones are not our ancestors, but we ourselves.

This book would not be worth the paper it is written on, if it only meant to make people aware of the spiritual dimension of our world. Its right of existence is rather due to the detailed description of this spiritual world, and of the quite evident consequences if we don't seriously deviate from our present course.

Strange enough, to the scientist the existence of a spiritual world becomes more evident, the better the scientist knows the physical laws. This curious phenomenon is described in the first appendix. In the second appendix four scientific proofs are discussed in favour of the existence of the unique human spirit. The last three appendices are philosophical essays on the Bell inequalities, on the nature of causality, and on the laws of spirit.

The five chapters of the book illustrate different forms of ingenuousness, which are a direct consequence of our decreasing ability to 'see' the spiritual dimension. There are plenty of movies illustrating how the evil one (the devil, for Christians) can get hold of a single person. The author of this book has no specific knowledge of exorcism, and diabolical possession is not the subject of this book. Rather, it focuses on ingenuousness, which could be described (with some imagination and with my sincere apologies to the theologians), as the diabolic possession of society as a whole. Ingenuousness is the kind of "possession" that Jesus speaks about when he commands Peter to back off ("vade retro, satana"), while Peter only tried to protect the Lord's life. It's the same kind of possession of young girls who believe they will find true love by clothing sexy, or of the clergy when they start messing with business and politics: for the people of this world are shrewder in dealing with their own. (Luke 16:8). The second chapter is devoted to ingenuousness of philosophers who think they can say anything about our world without knowing what sciences have to say about it. The third chapter discusses legislative ingenuousness, which is often due a philosophical misunderstanding of law, specifically, leading to excessive moralism which can be easily avoided by means of Hume's guillotine.

The fourth chapter handles the toughest nut, the inside jobs, because that touches directly upon the ingenuousness of the self-conceited adult. This chapter can only be digested by readers with a strong stomach and not too lazy to look up some of the references on the web. It might dramatically change your view on the world. The fifth and last chapter argues that ideologies should be treated as nasty relics of the 20th century. If our parliaments continue to discuss ideologies instead of plain numbers, our occidental societies will be no more.

Dedication

This book is dedicated to all people
(Jews, Christians, Muslims, people who believe
in multiple gods, or agnostics)
who wish to fight for peace in our world.
The hardest part is to know our enemy:
this is possible only for those who have thrown
off the shackles of ingenuousness.

Chapter 1: Personal ingenuousness

Occidental societies tend to ingenuousness. Nothing strange here, Jesus predicted it two thousand years ago. In this chapter a few examples of well-known ingenuousness are presented in order to get the reader acquainted with the subject. For the Christians among you: although our Lord said that *often* children of this world are shrewder than those of the Light, He did not say *always*. We should at least have the courage to open our minds to the possibility of being ingenuous.

Since it is always easiest to recognize defects in other people, let us start out, in this chapter, by discussing the easiest and best-known expressions of ingenuousness. Everybody knows the "mechanism" of how ingenuousness propagates:

(i) Parents like to leave in inheritance to their progeny a good, attractive and just world. They will try to avoid talking to their children about atrocities and injustices, out of fear of hurting them psychologically.

(ii) It is always difficult to admit that the world is much worse than it seems, even more so to the degree that we carry part of the guilt ourselves, by action and omission.

(iii) It is more comfortable to believe all that is fed to us by the media than to painfully evaluate the trustworthiness of the sources. The reader might ask: "Even in the case of BBC?" Yes even there. Chapter four gives a nice example of how BBC structurally lied and deceived in order to protect a source they were morally compelled to reveal.

The first section is dedicated to the ingenuousness of the girl (1.1) and the second to that of the clergy (1.2).

1.1 Ingenuousness of the girl

Girls between 12 and 15 like to show their corporal beauty. If they only knew what type of men their beauty attracts, and more importantly, that even fine young men lose their capacity to love when their sexual desires are easily fulfilled, these girls would spontaneously put on a burka. Apparently, not many parents have the courage to explain to their daughters the difference between 'being desired for one's potential to sexually satisfy a man' and 'to be loved by a man'. If nobody explains this difference to them, how are they supposed to find out? When they return home crying, because they were raped or because they feel as such, it is too late. The girl paid a heavy price for her ingenuousness and the parents have to live with their consciences.

It is better to prevent than to console afterwards. For smart girls a sincere and personal talk with either parent will satisfy. The simpler girls probably need something more visual. An offensive but clear example is watching "The Ugly Truth" (starring Gerald Butler and Katherine Heigl) with daddy. Daddy's task is to interrupt the movie every five minutes, to explain what exactly excites men sexually in the scenes; and to explicitly comment on how in sick minds sexual excitation is amplified by the humiliation and the pain of the woman; and how sexual consumption sickens the mind. These last ingredients are particularly revealing. It smells of the devil. Hence it can hardly be a surprise that practical materialism (which forgets about spiritual beings) is a highway to this kind of ingenuousness.

1.2 Ingenuousness of the clergy

This section treats two examples of Catholic ingenuousness: one from Rome, the other from Turin.

In 1940 Pius XII ordered an archaeological excavation with the end of studying what was underneath the principal altar of St. Peter's Basilica. Due to the German occupation they had to work secretly. Had the Germans found out, they certainly would have taken control of it. The Nazis were interested in all sources of power, be it technological or of a mental-mystical nature. The excavations were concluded in 1949. Pius XII announced, through a radio message on Christmas 1950 that the tomb of

Peter, the first Pope, had been found a few meters right below the principal altar of the basilica.

The Italian archaeologist and professor of Epigraphy and Ancient Greece, Margherita Guarducci testifies, "The actual altar (of Clement VIII, dating from 1594) is built on top of that of Calisto II (1123); that was on its turn built around the altar of Gregory the Great (590-604); this last one was built on top of the so-called "monument of Constantine" (312-326) which contains the first monument of Peter, even older, which goes back to the second century. (…) Inside Peter's monument is found part of a small building attached to a certain red wall which was the background of the first monument of Peter. In the interior of the building there is a wall filled with graffiti (…) dating before Constantine's monument (…). The density of graffiti on the wall testifies to the devotion of the faithful. The first monument of St. Peter has a lid on the floor, covering an ancient tomb (…)".

In 1952 Guarducci asked and obtained permission by Pius XII to study the graffiti because of an inaccuracy in an archaeological publication by Antonio Ferrua, S.I. Guarducci directly started to look for the rock that had the graffiti 'ΠΕΤΡΟΣ ΕΝΙ' (with the sense of 'enesti': Peter is here inside). She did not find it: Father Ferrua had taken it home.

According to the multi-secular tradition of the Church, Peter's bones had been placed in the tomb under the lid. Why did Ferrua and his team of archaeologists failed to find them? Due to pure incompetence: they used a rough instrument, similar to a gardening shovel, but with a long iron handle, to create cylindrical holes in hard soil ('cartoccia' in Italian) wherein to fix poles. To get to Peter's tomb without too many circumventions, the excavators crashed Calisto II's altar without hesitation. Under the pounding of the cartoccia, Peter's tomb, which initially contained only his bones, was filled with debris. Having reached the designated spot, to their surprise the excavators only found debris where they possibly had expected something spectacular, in the style of a nicely polished arch of the covenant of solid gold (Indiana Jones had his less fortunate predecessors). So they decided to continue their search in nearby sites, getting rid of all the debris straightaway. Monseigneur Ludwig Kaas, the supervisor responsible for the excavation (who did not trust his four genius archaeologists very much), had the habit of inspecting all "debris" at the end of the day. He also did so in this occasion, together with the "sanpietrino" (helper in the basilica) Giovanni Segoni. One of the two noticed something protruding from the debris, which looked like a human bone. Out of mere mercy for the dead (neither Monseigneur Kaas nor Segoni had the slightest idea of holding a bone of St. Peter),[1] they collected all other bones in the debris, catalogued and archived them as originating from the so-called G-wall. Ten years later, when Guarducci was taking a coffee with Segoni and interrogated him about those inspections with Mgr. Kaas, he remembered having collected a set of human bones from the G-Wall... A year later (1968) Pope Paul VI had to declare that not only was the tomb of St. Peter found, but also his bones, exactly under the main altar of St. Peter's Basilica – not after having been reprimanded severely by Guarducci with regard to the relics.

The second example of baffling ingenuousness of Catholic clergymen concerns Cardinal Anastasio Alberto Ballestrero O.C.D., archbishop of Turin and pontifical guardian of the shroud. The Spanish Wikipedia mentions: "In 1988, the Holy See permitted three centres of investigation to carry out independent radiocarbon tests on a piece taken from the corner of the shroud. The piece to extract was selected meticulously by textile experts, Professors F. Testore from the Department of Material Science (Polytechnic University of Turin) and G. Vial from the Museum of Textiles (the International Centre

---

[1]   Wikipedia recounts the conclusions of the Italian anthropologist, Dr. Venerando Correnti:

- The bones of the animal (a mouse that did not find a way out of the tomb) are practically clean compared to human remains, because they were covered with dirt that was later analysed and found to be from the open and empty tomb which was identified as St. Peter's tomb (because of the inscription 'ΠΕΤΡΟΣ ΕΝΙ', notes the author). On the other hand, all the other tombs next to this finding have a different type of dirt (in addition to inscriptions of the kind "Peter, pray for us, who are buried next to your body".).
- The bones have a reddish colour. This is probably due to a purple cloth in which they were wrapped. The remains of gold thread confirm that these were the bones of some venerated person. It is possible that the bones were taken from the original tomb to 'store' them in the niche so they would be kept safe, for the niche was intact starting from the time of Constantine until the findings.
- The bones found here belong to single person: strong, male, of advanced age (around seventy years old), and from the first century.

for the Study of Ancient Textiles) in Lyon, under the supervision of Michael Tite, Head of the Research Laboratory in the British Museum. Three samples from the shroud were sent to three different laboratories, Oxford University, University of Arizona, and the Federal Polytechnic School of Zurich, together with three samples of fabric from an Egyptian tomb dated 1100 BC; from a mummy from the year 200 BC; and from the Luis IX's cape, from the 13th century. The laboratories did not know where each of the samples came from. In this way, the reliability of each technique could be verified, whatever the sample age. The three laboratories dated the control samples and they coincided in dating the "shroud fabric" between the 13th and 14th centuries (1260-1390). In 2002, Ray Rogers, an expert in chemistry from the STURP team and a retired member of the National Laboratory of Los Alamos, postulated that the sample cut from the Shroud of Turin in 1988 was taken from an area of the linen that had been re-knitted during the Modern Age."

The ingenuous author of the Wikipedia page apparently believes Roger's thesis, not any less ingenuous. They both fail to mention the necessary consequence of this thesis, namely, that professors F. Testore and G. Vial are utterly incompetent. Well, there is no reason to believe such accusation, considering the long list of scientific publications in prestigious journals from both scientists. Is it really so difficult to believe that Michael Tite simply cheated by exchanging samples? He was the only and exclusive supervisor, without any form of control! Moreover, prior to the investigation he had already officially declared that the shroud could be nothing but a fraud.

Tite's cheat[2] will not be proven in detail here. Core of the argument is that the sample sizes as taken from the shroud by Testore and Vial do not match with any of the sample sizes delivered to the three laboratories. My best guess is that Michael Tite lit his cigars with the shroud samples.

Cardinal Ballestrero believed to make a 'bella figura', as the Italians say, by getting rid of all the Catholic teams, and leaving the direction in the sole hands of a declared atheist. Imagine the opposite: that the

---

[2]   https://www.ewtn.com/library/issues/sturp.txt

only supervisor would have been a practicing Catholic, who prior to the investigation would have declared his absolute faith in the authenticity of the shroud, and that, oops, indeed the shroud was measured to be 2000 years old! Not even Catholics would have believed such a story! Therefore a solid protocol is of utmost importance.[3]

---

[3] There is no need for a large financial investment to realize a cooperation of multiple supervisors, each with only part of the information. In order to prevent any communication of the supervisors with the laboratories after having handed out the samples, the supervisors should be physically isolated until all the measurements are carried out. Whoever is not willing to publish a protocol and test it in public trial is unprofessional. If in addition all Catholic teams were disregarded (Ballestrero even excluded Joseph Ratzinger, in spite of his begging explicitly) unprofessionalism is gracefully married to ingenuousness. Bravo Ballestrero!

Chapter 2: Philosophical ingenuousness

The terms ingenuousness and naivety are not distinguished in this book. There is a distinction, however, between ingenuousness in behaviour with merely private consequences, and that with social consequences, and the context should make it clear what kind of ingenuousness is being discussed. Of course, all peoples are different, and they all have their specific nuances of ingenuousness. The Dutch writer of this book would not even know to what extent the British examples of ingenuousness are typically British.

In this chapter philosophical ingenuousness will be studied. I regret to say this so crudely, as a physicist to philosophers, but the fact is that very little of what philosophers have said in history, still stands today. The reason is simple: philosophy is a more difficult discipline than natural science. Some reader might object: "But all scientific advance is possible only by means of logic, and that is a philosophical discipline". This view the author does not compart: logic could also be claimed by mathematicians, and even by housewives doing their daily shopping.

2.1 Philosophy and science

Generally feared is Stephen Hawking's thesis: 'philosophy is dead'. Although I agree with Hawking's reasoning, I do not endorse his conclusion. The reason for this nuance (between the conclusion and the reasons leading to it) is in the definition of the term 'philosophy'. For Hawking, philosophy is nothing more than an 'ancilla scientiae' (slave of science), since according to him matter is the only thing that exists. As explained in full detail in the first two appendices, such belief is erroneous; thus it is not very useful to study Hawking's concept of philosophy.

Physical discovery has always preceded philosophical thinking: most clearly in the case of momentum conservation discovered by Buridan[4] and Oresme[5] in the 14th century; but also quite known are the inexistence of the electrodynamic ether; the existence of a quantum-mechanical wave function for every possible physical system, nay even of the universe; the spatio-temporal relativity of Einstein, and so forth. In all these cases the role of philosophy has been at best to hamper physical progress due to the philosophers' anti-reactionary clinging to Aristotle and Plato.

---

[4] Jean Buridan (1295 – 1363) was a French priest who sowed the seeds of the Copernican revolution in Europe. He developed the concept of impetus, the first step toward the modern concept of inertia, and an important development in the history of medieval science. His name is most familiar through the thought experiment known as Buridan's ass (a thought experiment which does not appear in his extant writings). Born, most probably, in Béthune, France, Buridan studied and later taught at the University of Paris. Unusually, he spent his academic life in the faculty of arts, rather than obtaining the doctorate in Theology that typically prepared the way for a career in Philosophy. He further maintained his intellectual independence by remaining a secular clerk, rather than joining a religious Order. By 1340, his confidence had grown sufficiently for him to launch an attack on his predecessor, William of Ockham. Buridan also wrote on solutions to paradoxes such as the liar paradox. An ordinance of Louis XI, in 1473, directed against the nominalists, prohibited the reading of his works. The bishop Albert of Saxony, himself renowned as a logician, was among the most notable of his students.

[5] Nicole Oresme was born around 1320 in the vicinity of Caen, Normandy. The fact that Oresme attended the royally sponsored and subsidized College of Navarre, an institution for students too poor to pay their expenses while studying at the University of Paris, makes it probable that he came from a peasant family. Oresme studied arts in Paris, together with Jean Buridan (the so-called founder of the French school of natural philosophy), Albert of Saxony and perhaps Marsilius of Inghen, and there received the Magister Artium. He was already a regent master in arts by 1342, during the crisis over William of Ockham's natural philosophy. In 1348, he was a student of theology in Paris. In 1356, he received his doctorate and in the same year he became grand master (grand-maître) of the College of Navarre. Around 1369, he began a series of translations of Aristotelian works at the request of Charles V, into the French language rather than the (at that time) more popular Latin. Oresme's works have been published much more in Britain than in France (exception made for the great Pierre Duhem), because Oresme defended the continuity across the middle Ages and Modernity.

There simply exists no historical example of philosophical help to scientific advance. Quite the contrary, examples abound in which philosophers joined all forces to fight against new physical models. Aristotelianism easily survived the defying teachings of Buridan. Aristotelianism even managed to ban Galileo Galilei in the 17th century, that is, 21 centuries after the master himself! The downplay or opposition of philosophers to scientific innovation is mainly due to the fact that philosophers tend to think *a priori*, while natural scientists think *a posteriori*: they start observing, then define their quantities, and if possible, a crucial experiment, and only then try to come up with a mathematical model that describes all the results. For a natural scientist, this is the meaning of 'understanding nature'. That is where the philosopher's task should begin, really. No doubt Aristotle is the greatest philosopher of all times. His major contribution to philosophy is known as hylomorfism, which explains how an object can change without losing its identity. Applications of hylomorfism will be given in the third and fourth appendices.

However, from however many historical instances it cannot be deduced that natural science will keep preceding philosophy also in the future.

2.2 Behe's Intelligent Design

The most popular attacks on faith come from atheists of renowned scientific importance (Daniel Wegner, Daniel Dennett, Ian Glynn, Edward Wilson, Richard Dawkins, Stephen Hawking) with erroneous philosophical opinions. To these attacks Christian philosophers generally respond with purely philosophical arguments. To scientist such arguments sound circular, as if the conclusion were already included in the premises.[6]

---

[6] For example, when philosophers argue the impossibility of evolution because it implies that the complex and the active spring from the simple and passive, scientists answer that the conclusion is already contained within the premises. Everybody knows that a program can have as many outputs as there are inputs. Now consider the first organic cell as a genetic program. Depending on the circumstances (the input) all types of output are possible, even an evolutionary path toward the human body. Did the complex proceed from the simple? No, the evolutionary path toward the human body was already contained in the genetic programming of the first cell.

Christian philosophers don't want scientists like Michael Behe to take a philosophical stance at all. They hypocritically confuse 'intelligent design' with creationism.

The 'strong' creationism idea teaches that the earth was created by God six thousand years ago, because that is what the Old Testament describes, when the ages and the relation of kinship are interpreted literally. They have no problem in declaring that God created the world with the all its paleontological sites and fossils along. Of course there is no way to prove them wrong. Nay, it's equally possible that God created the world yesterday and in doing so, he filled our minds with the pertinent memories. Another option is that there is no material world at all, but everybody dreams.

Science is not about gathering all options that *could have occurred*, but selecting that only option *which is the most likely to have occurred, assuming that God does not fool us*. Strong creationism is incompatible with this view on science: these are people that seem to revel in a God fooling us, the more the better.

'Weak' creationism accepts the age of the universe (around 13.7 billion years) and of the earth (around 4.5 billion years), but it dogmatically refuses to accept the evolution of species. For them the stability of species is a perpetual miracle. The comparison made mostly by Italian and Spanish philosophers between Behe's 'intelligent design' and creationism is scientifically incorrect. Michael Behe is an American biochemist who wrote a challenging book in 1996, called *Darwin's Black Box*. In this book, he unmasks the pseudo-scientific positions of well-known evolutionists. He also defends the impossibility to explain, on the basis of only Darwinist principles (variation and selection), the evolution of biological structures like bacterial flagella, cilia, or biological processes like the cascade of blood coagulation, the immune system, or vesicular traffic. His argument is based on the concept of 'irreducible complexity'. For example, a mouse trap consists of a few basic elements. A watch has a few more elements. The complexity of an object with a certain function is irreducible when, taking away from the object whatever piece, its function is compromised, even if we were to rearrange the remaining pieces. Behe makes it clear that the Darwinist principles (variation and selection) can only operate on biological constructs that succeed in carrying out a function. The constructs that do not succeed can never win the evolutionary race.

Only the 'working' constructs can run the evolutionary race. The better they perform, the more they multiply, and the more their genes abound, resulting in a greater possibility of surviving the evolution.

Do not get me wrong: I am not saying here, nor does Behe say, that Darwinist principles (variation and selection) are invalid or erroneous. They are perfectly valid and there are many examples proving their validity. A beautiful example of the role of variation and selection is presented in appendix A2.2, where it is discussed how the sterility of Hymenoptera worker bees could originate 11 times independently in the history of evolution. This example also makes it clear that the fundamental player in evolution is not the individual, nor the species, but the (group of) genes, or even better said: *genetic information.* Genetic information does not care if it multiples in a bear or in a bear parasite; in nucleic acids or in amino acids. But let us come back to the problem of the first complete, operating cell; what Behe says and every healthy scientist with him, is that the Darwinist principles can modify existing species, but it is not yet established how they ever could *generate them.*

Behe asks how it is possible to produce the necessary intelligence to make a mouse trap, or a watch, by means of only variation and natural selection, if there is not any one on the outside that decides which programmatic variations will approach us to the goal, and which ones will lead us away. The answer to this question is unknown today, but it is a valid scientific question, whence there exists a valid scientific answer. In his 'On the origin of species' Darwin confesses that he has no idea how his theory could ever explain the evolution of an organ as complex as an eye. Each biological attempt to develop a new organ sacrifices robustness: if not, eyes would spring up continually all over our body. Without a connected cerebral circuit, the eye would not be useful. The same goes for the evolution of the cerebral circuit: without eyes, the cerebral circuit for vision is useless. Naturally it is possible that an organism develops an eye and a well-connected visual circuit at the same time, although it sounds a bit like a miracle…

Granted, with his book, *Behe did not add anything to the positive scientific knowledge;* concretely, he did not resolve the basic problem (the evolution of the first cell), nor did he offer any specific suggestions. But such was never his intention — he only leaves it clear that those who finally come up with the solution will certainly not be hard-core Darwinists, because they do not even see the problem. For them everything is already explained and understood in terms of the Darwinist principles of variation and selection. Excellent! This only means less competition for the next biology Nobel Prize.

Let us now move on to the ingenuousness of the ethologist and evolutionary biologist, Richard Dawkins (thank God that we Christians are not the only ingenuous). Toward the end of the past century, Dawkins thought that he could mitigate the exuberant statistical improbabilities by subdividing the evolutionary problem into distinct 'phases'.

If the gene carrying the essential information for an evolutionary leap would have to change on 120 sites (refer to appendix A2.2 for terminology), Dawkins would define 120 'phases': one phase for each genetic mutation. Every genetic site is occupied by one of the four different nitrogenous bases: Adenine, Cytosine, Guanine, and Thymine (for DNA) or Uracil (for RNA). If each variation affected all 120 sites with equal probability (the most reasonable supposition), the probability of making the evolutionary leap is one in 4^{120} per mutation. If the mutations occur each second, there would be an average waiting time of 4^{120} seconds, which is of the order of 10^{65} years. The earth is 4.5 billion years old. For one evolutionary leap to occur in the full lifetime of the earth, a constant population of 10^{55} individuals is needed. Well, the earth does not even contain that number of atoms. Therefore Dawkins proposed his theory of phases: mutations only affect those sites that do not yet coincide with the wanted one. In this case a similar calculation teaches us that the evolutionary time needed for yielding the evolved species is only a few minutes, rather than 10^{65} years.

Behe points out that Dawkins' theory of 'phases', instead of being Darwinist, is blatantly creationist: indeed, it supposes an intelligent being that, from the outside, determines at every moment which sites mutate and which not. Behe concludes that Dawkins' model denies the very foundation of Darwinism. Dawkins did not realize that his model of 'phases' needs a God acting as the final cause.

That is quite interesting. When Dawkins launches an openly creationist theory, nobody complains, apart from an unknown biochemist. When Behe replies with the introduction of an interesting and new concept (irreducible complexity) the evolutionists are furiously indignant. They dub him a creationist and a choir of retarded Christian philosophers join the club.

For Dawkins, it is a bitter irony of history that his creationist model has an important ingredient for the scientific explanation of the creation of man. To appreciate this, Dawkins' library needs two more notions: quantum mechanics and entropy.

The notion of entropy is treated elegantly by sir Roger Penrose in his classical book *The Emperor's New Mind.* Among others, Penrose calculates the enormous difference between the initial entropy (of the universe at the moment of the Big Bang) and that of our present universe. This difference is so fantastic that Penrose chose to illustrate the Big Bang with a magician swinging his wand, short of an explicit religious confession.

The smaller the initial entropy, the larger is the amount of macroscopically different final states. A room uniformly filled with air has a very high entropy, and very little can be expected to change the fact that the room is uniformly filled with air. The situation would be quite different if all air molecules were concentrated in a light bulb, with no air in the rest of the room. Then the entropy would be very small, and many final states could occur (due to the explosion of the bulb). An equivalent argument holds for our universe, although it is slightly altered by the existence of gravity. Here it suffices to summarize that it is exactly the small initial entropy that makes many final states possible: even states with biological life (although the number of these states is amazingly small). Finally, all universes, with or without life, die an "entropic" death: they will end up as a collection of black holes in an ever expanding universe.

On the other hand it is generally believed since 1964, when John Bell wrote down his famous inequalities (see appendix A3), and generally accepted since October 2015, when Hanson performed a loophole-free Bell experiment, that quantum choices do not issue from the arrangement of matter. The low initial entropy of our universe requires a huge amount of quantum decisions by a non-material entity (let's call it God, in order not to multiply spiritual beings unnecessarily) to arrive at its present situation: 13.7 billion years of continuous quantum choices, or 'directed evolution', which is a different word for the same thing.

Putting together the notions of Behe, Penrose, Bell and Dawkins, a model for the creation of man results, which is compatible both with the Catholic Magisterium and with science (quantum mechanics as well as gravity). Traditional Christians should conform to the human body having evolved along a sequence of fantastic creatures, last of which the common ancestor with the chimpanzee. They should equally conform to the fact that the fertilised human ovum (egg cell) is nothing but an animal, until God wishes to breathe the human spirit into it. And nobody, no scientist, no angel, not even the Holy Roman Catholic Church in all its infallibility, can ever predict at what moment God decides to breathe the human spirit into a fertilised human ovum: the beginning of spirited human life will always be an explicitly divine, unique, personal, irrevocable and unrepeatable act of creation. We humans should consider this as an honour: clearly, God does not consider us as pieces in series, but treats every single one of us as if we were the only creature in the universe. The material aspect of the human body may have evolved along a pre-established plan, in agreement with ruthless laws of nature, but the human spirit came to being instantaneously and by the direct will of God.

The evolution of species and especially the human body is the result of a series of 'directed events' (quantum elections) throughout the history of evolution; all of this without the least contradiction with physical laws. With exactly the same initial conditions of our Big Bang, but with an extremely slight variation of quantum elections throughout history, there would have been no life at all in our universe.

The only way to confute this 'Weltanschauung' consists in finding life elsewhere in our universe. Until today our radios only detected background noise; extra-terrestrial life does not yet make it beyond fiction movies. If the author of this book is right, no life will ever be found elsewhere in the universe; not because of a philosophical impossibility, but because of physical improbability.

2.3 Hume's Guillotine

Hume's guillotine is the elimination of 'is-ought' errors. An 'is-ought' error is the illicit identification of natural properties with moral properties; in simple words, the illicit substitution of the verb 'is' with the verb 'ought'. An example of an 'is-ought' error would be to consider every religion good. This is an instance of a more general philosophical error in which act is mistaken for potency, or spirit for matter. Both in these pairs, as in the pair of moral and natural properties, the terms relate like act to potency. Evidently, the identification of act and potency is a serious philosophical error. In this section we will apply Hume's guillotine to two similar concepts: to democracy and to democratic laws.

David Hume was a Scottish philosopher from the 18th century. In his book 'A Treatise of Human Nature' he graciously complains "that his colleagues tend to start their papers in a descriptive and deductive way, up to a certain point; then, almost imperceptibly, the verb 'is' disappears, and instead the verb 'ought' appears in its place." A posterior (20th century) philosopher, George Moore, coined the concept 'naturalistic fallacy' to indicate the impossibility of defining moral categories. Even though both of them commit a crucial philosophical error (see appendix A5.1), Hume was right in his critique of Christian philosophers.

A democracy is established when an independent country decides that their governors be freely elected in an electoral campaign with more than a single political party, by popular vote, counting each vote equally. This is not meant as a full-blown definition of democracy, but rather to highlight the absence of any moral notion. Many Christian writers agree theoretically, but not in practice. For them, democracy is not realized unless moral. However, there should be no difference whatsoever between democracy and tyranny of the majority. Then, application of Hume's guillotine to the concept of 'democracy' (literally 'power to the people' in Greek) has the following consequences:

(i) democracy has nothing to do with morality; there can exist and there have existed moral and immoral democracies; the same goes for monarchies;

(ii) the responsibility of granting freedom and tranquillity in society is not inherent in the concept of democracy, but it is derived from the concept of morality.

In practice it is impossible to measure the morality of a society's government: we would have to sum up all of the injustices suffered by each citizen at the hand of civil or military authorities and at the hand of other citizens. There is no other option than to look for indicators of morality of a government. *We argue here that the most efficient indicator of morality of a government is economic growth, because economic growth requires all citizens* to exercise human virtues at all levels; rich and poor, high dignitaries and modest employees. The economy is like a clockwork process in which all parts are optimally integrated. Of course the morality of the people is to be distinguished from that of the government. If the government is corrupt, there will be little if any economic growth, however virtuous the ordinary people. Hence the indicator of economic growth does not represent the average morality of the people, but a severely weighted average: weighted by the influence people have on decisions taken by the government.

Naturally, economic growth is not an indicator of instantaneous morality, but of that over a certain past period, usually with a delay of a few years. Sometimes there is a delay of many years because it takes time for governmental decisions to affect the economy. Below follow three examples of economic consequences of immoral decisions.

- Chinese communism fixed the maximum of one child per couple. This measure powerfully pushes economic growth at short term (specifically when a second one is admitted against a monetary fine): it eliminates the poor, rural population (the principle defiler of statistics) and it encourages the growth of the population in rich areas. Unfortunately the Chinese ideologists did not foresee the nasty consequence of their immoral law: that the poor preferably aborted their females! The social disaster is indescribable and the pernicious economic consequences are unavoidable. Chinese stocks are already falling.

- In Greece they had it all, great thinkers like Aristotle and Plato, cultural treasures (Romans loved Greek art) as well as natural (beaches, mountains, valleys); but a long tradition of corruption has led the country to the border of bankruptcy. As long as the high political dignitaries continue being corrupt, nothing will ever change; and the more money Europe invests in them, the more corruption will prosper.

- Germany, the impressive economic engine of Europe, is paying for their anti-nuclear phobia, which is pure irrational sentiment; comparable with the fear of the English farmers of the 19th century that saw in their fields this metallic monster on wheels, driving the livestock mad (bitter milk, biological abnormalities, and many more horrors) with its sounds and smoke plumes; a monster that city folks called 'a train'. If they do not start to reason the Germans will lose much economic power. Exactly the same holds for my own country, The Netherlands.

Another example typical example of the 'is-ought' error is that of the Christian lawyers who believe that the goal of human laws is safeguarding morality. This is obviously false. Such objects already exist, and they are called divine laws, e.g., Moses' Ten Commandments. Similarly, some Christian lawyers claim that the goal of imprisonment is punishment. This is equally false. The institute where punishment is inflicted already exists, and is called purgatory. These are two examples of illicit 'divinization' of earthly stuff. 'Divinization' is the act of attributing supernatural properties (divine punishment, divine law) to merely natural concepts (prison, human law). The application of Hume's guillotine comes down to removing the inexistent supernatural character:

(i)   The primary goal of human laws is to safeguard liberty and tranquillity in society.
(ii) The direct goal of prison is twofold: to prepare benevolent convicts for society, and to separate malevolent convicts from society.
(iii) The goal of a sentence is not punishment, but avoidance of vengeance, and re-establishment of commutative justice as far as possible.

The goal of punishment is of no consequence, for punishment does not belong to the order of human justice.

2.4 Legislative divinization

The principle cause of unjust occidental laws is the ingenuousness of the legislator. As it was mentioned in the previous section, Christian lawyers tend towards divinizing law. If in some cases the public is angered due to an unjust verdict, those lawyers would say that the public is not aware of the complexity and intricacy of law.

Laws are there for humans, and not the other way around; if this already applies to divine law, how much more should it apply to human law? If the public is angered, it is an unequivocal sign that something is not working (not necessarily the wording of the law — its enforcing can also fail). In any case, blaming the public's ignorance reveals the failure to grasp the essential: laws are the medium by which peace and liberty in society are secured. Even though the law might be a good one in terms of whatever moral doctrine, if its implementation provokes social unrest, it is a bad law. *This means by definition that good laws are different in different countries.*

In my country it happens regularly that the public reacts indignantly with respect to a verdict. A good Dutch example is that of an armed robbery in a jewellery store in a town called Deurne in 2014. When a woman noticed on the video that two armed men were threatening her husband in the store, she took her car right away to the jewellery store and killed both of the assailants that had the husband at gunpoint. What do the mentally retarded officials from the Dutch public ministry do? They ask for 200 hours of task punishment plus half a year of conditional imprisonment *for keeping a loaded gun in his store without a license.* The poor man decided to buy the gun after suffering a previous armed robbery, which happened four years earlier. No need to say, nobody us interested in monetary compensation of the jeweller, for financial losses and for psychological damage.

Examples like this abound in other countries. For example, the state prosecution in Jacksonville, Florida, which threatened Marissa Alexander (34 years old) with 60 years of prison for having fired a warning shot in her own house to end the harassment by her ex-husband. Under the threat of prosecution Alexander admitted to three charges of aggression, which yielded her 20 years of imprisonment. To defend herself against the prosecution's injustice, Alexander only had to call her two sons to testify — something she chose not to do because the prosecution would have torn them (and whatever little was left of the family bonds) apart in the court of law. Alexander was sentenced and imprisoned. It was the strong civic pressure that finally liberated her, after three years of prison.

The ex-husband, on the other hand, did not feel such restraint — he called in his 15 year old son (not hers), who was in the house when Alexander fired her gun. The boy declared that 'his life changed the day that his step-mother shot a weapon in his presence'. He did not say that 'he had the scare of his life' but 'that his life changed; it is not difficult to think of the lawyers feeding the boy whatever answers the jury needs to hear. Imagine a 15 year old boy ending his declaration with the words, "I was not physically hurt, but I was emotionally and mentally injured."

The most convincing argument against the lack of justice in criminal law is the great difference between states. In the United States, self-defence against the intruder differs from the 'castle doctrine' at one end of the spectrum, 'stand your ground' half way, and the 'duty of retreat' at the other end. In the first case, there is no need to assess the intruder's intentions, nor of the lethality of his weaponry. In the second case, the owner has the right to keep his position using a proportional amount of self-defence, which does imply the duty to assess the combat ability of the intruder. In the third case, there is an obligation to retreat. I don't know which of these is the most just — what I do know is that all three cannot be just, for being incompatible, because the local circumstances differ very little. The sad truth is that many laws have been passed without any previous research; they only try to combine historical heritage with recent cases.

Chapter 3: Legislative ingenuousness

3.1 Talion and composition

We have just seen that the justice or quality of a law is given by its contribution to social peace and freedom. The ancient Jews did not have the slightest inconvenience in determining the measure of mending in case of deliberately inflicted damage: "an eye for an eye, a tooth for a tooth", says the Talion (Ex 21:24). Historically, this law was established to end the hereditary resentments and vengeance, whenever the "Composition" failed: the private payment of a sum of money agreed upon by the aggressor and the victim or the victim's family.

The Talion consists in making the delinquent suffer what his victim suffered. The Talion is the first historical form of punishment that assumes the existence of a public power that applies a material equivalent between the damage suffered by the victim and the damage inflicted by the aggressor. The Talion succeeds in preventing vengeance, but fails in compensating the victim.

The Composition is an indexation of the harm done, attributing a price to every crime. Germanic tribes of the 12th century used the term 'weregild' (from the Latin 'vir', man, and the Saxon 'geld', money) for the price of a homicide. Among the Alemannic tribes the female weregild was twice the male's (assuming equal social rank). For the Saxons it was the other way around, more in line with contemporaneity. Much ahead of their times, these Alemannic!

The distinction between public delicts (crimes) and private ones was already known among the Romans in the 5th century B.C., and written down in "the laws of the XII Tables" (also called "of roman equality", or "decemviral") and contained norms regulating daily life of the roman people. Crimes were persecuted by the state, while all other delicts were treated privately. During the Republic and the Empire the private domain was gradually absorbed into the public domain.

During the Middle Ages (5th through 11th centuries) the three sources of law (roman, feudal, and Canon) mixed to some degree. These were times in which delict and sin were interchangeable, and likewise punishment and liberation. The end of the middle Ages is characterized by the rise of monarchies, where all power was concentrated in the hands of a monarch. The 12th century is best known for its reappraisal of Roman law. This was taught at several universities, initially mostly Italian, along with Germanic and Canon law. In the following centuries (13th through 16th) private retaliation was not tolerated any more, and lawyers generally agreed on the fact that the goal of punishment is intimidation, or discouragement. Degrees of juridical imputation are acknowledged: for the fool, the furious, the youngsters etc. The 17th century is associated with the flattering name 'enlightenment': Beccaria and Montesquieu introduced a series of penal reforms, humanizing the punishments, ending torture, introducing equality before the law, the principle of legality, and proportionality between delict and punishment. The Italian positivism introduces a new concept in the 18th century: the punishment is determined by the probability of recidivism (Lombroso, Ferri), rather than by the act. In the 19th century Neo-Kantianism criticizes the positivism for lacking scientific base. In the course of penal history the goal of punishment has gradually shifted from a religious to a consequentialist perspective. The latter perspective is embraced by the author of this book. It is proposed that punishment is determined only with an eye on protecting liberty and tranquillity in society. The application of justice might be a consequence of defending tranquillity in society, but not the means. An important advantage is that it facilitates the cooperation of people with different beliefs and moral systems.

3.2 Nocturnal Intrusion

The criminal laws of a country are too important to leave them to historic arbitrariness, or to the free discretion of a few lawyers and judges, even if they are good Christians. Laws require interdisciplinary, profound, quantitative study of all relevant correlations. Let us present a simple example. How many instances does the reader of my book know of silent nocturnal intrusions by a tourist asking for directions? Yes indeed, silent nocturnal intrusions are practically always malevolent. Then, it is not a waste of time, but also an injustice to demand that the victim should prove anything about the intruder: neither intention, nor level of self-defence.

Some three thousand years ago the Ancient Jews already used the diurnal criterion as a decisive circumstance. The Exodus makes it clear that whoever kills a nocturnal intruder is not guilty: this sounds much like the "castle doctrine" mentioned in section 2.4. The first verses of Exodus, chapter 22 read, "If the thief is caught while breaking in and is struck so that he dies, there will be no blood-guiltiness on his account. But if the sun has risen on him, there will be blood-guiltiness on his account. He shall surely make restitution; if he owns nothing, then he shall be sold for his theft."

It is not at all proposed here to base a modern juridical system on Exodus. But for the self-conceited Christian it is important to remember that when Jesus said that he did not come to abolish the law, but to fulfil it (Mt 5:17-19), he was most specifically referring to the Ten Commandments and a little less specifically to the four legislative chapters (20-23) of Exodus. Yet it is understandable that Christians are somewhat puzzled by the enormous differences between the new and old laws. The difference between the two is that the old law seeks justice, and the new one charity. Jesus proclaimed that whoever gets his right cheek slapped should present his left. Does Jesus' new law contradict the old law, which states that an intruder should be thrown out with violence? Of course not!

What would Jesus want a poor family father to do, when his exuberantly rich neighbour requires his only sheep? Possibly offer his only ox as well, and resign to contemplating the slow hunger death of his wife and children?

The idea is quite simple. The old law seeks justice, and Christians are required to apply them whenever the rights of defenceless people are trodden. As soon as it comes to one's own rights, one has the choice: apply the old law, and stay put in the love of God, or apply the new law, and grow.

On the other hand, if nocturnal intrusions are noisy, the great majority of the cases deal with drunken friends. And how many more cases do my readers know of noisy nocturnal intrusions that ended with the accidental deaths of the friends by the landowner?

Lawyers might exclaim, "If we return to the castle doctrine, there will be an explosion of nocturnal intrusions with an unnecessary loss of many lives." Initially yes, because there will always be a misinformed thief. But soon enough thieves will change their strategy: they will not intrude during the night for the high risk involved. Bingo! That is where we wanted to arrive...

What really cries to the heavens is the contrast of communal funds invested in criminal and victim. The market value of a person's possessions is incomparable with its subjective value. For example, an old computer could yield 50 euros on the second-hand market, while its owner would gladly pay 500 euros for its recovery.

3.3 Self-defence

Another example of legislative-juridical naivety has to do with a sexual assault in Pamplona, Spain. A woman was harassed at 8:37 in the morning of the 'Sanfermines' in 2014 by a drunken man who wished to have sex with her. After participating in the traditional bull runs her boyfriend appeared on the scene. Upon seeing his girlfriend harassed, he ran towards her and punched the assailant.

The latter was taken to the hospital. The public prosecutor's claim amounted to half a year of prison and 120,000 euro fine for unnecessary violence on the part of the boyfriend, from which 60,000 euros to be given to the delinquent; and a year of prison for the assailant for sexual harassment with the *mitigating circumstance* of drunkenness.

Whoever understands the least of Hume's guillotine would never consider drunkenness as a mitigating circumstance, but to the contrary, as an aggravating circumstance. Ideologies have been proven unable again and again to change human psychology or the human heart. In every society there are both people susceptible to anger and docile people, both miserable people who need the bottle and those who don't. Hence it is useless to try to reduce viciousness in society. But drunkenness can be reduced by appropriate laws, so that is where laws have to act.

In addition, what is the message to society, provided by the public prosecutor? Pardon me the sarcasm.

"Come, heavy drinkers from every corner of the world, come to Pamplona! The more you drink, the less imputable you will be. If by chance you harass one of our beautiful girls, and a dumb local thinks he should come to her aid, our retrograde laws will have him pay all of your hospital expenses. With a bit of luck you might even meet him in prison and teach him a little lesson."

Another consequence of the loss of spirit is the idealization of human judgement. For materialists, human judgement is mechanical fact. For those who know the spiritual dimension of man too, it is clear that, from the moment the boyfriend sees her girlfriend being harassed, he is filled with anger and adrenaline, and loses his normal judgement. It might be difficult for the public prosecutor, sitting in his comfortable chair, to appreciate such a detail of human condition, but alas, that is part of the job; if he is unable to appreciate such nuances, he is clearly unfit for the job. So if the victim's boyfriend displays super-minimalistic behaviour, part of it should be pardoned because the assailant caused that state of mind.

Parents know very well that it is more difficult to adopt a comprehensive attitude toward their children than a severe one, but for the good of the children it is not always the better choice. Of course, the delinquent is pitiful and miserable, and the boyfriend probably not; but that is no reason for making them both pitiful and miserable.

Judges prefer to accommodate the delinquent and disappoint the victim, especially if the assailant is powerful and the victim harmless. A criminal could take revenge (even from prison), while nobody fears the harmless righteous citizen. How comfortable!

As far as the way being personally attacked influences one's judgement, think about the case of a Costa Rican goalie, Esteban Alvarado Brown playing in a first division Dutch football team. In 2001, he was assaulted *during an official football game* by an Ajax fan that ran into the field and threw him a karate kick. Alvarado was barely able to avoid the kick and he responded. Alvarado went super-minimalistic: he continued kicking the assailant after he had fallen to the ground. Alvarado did so instinctively, even though the whole stadium (and half the country via television) was witnessing the events. The cowardly referee gave Alvarado a red card, according to the booklet. A total disaster on all accounts. Alvarado's coach had to withdraw all his players from the field, thereby losing the match, in order to avoid collecting eleven red cards. The frustrated public had paid to watch a football match, but ended up watching a poor karate exhibition.

Of course the legislators aim at discouraging violence on the football field. But is this the way to do it? Like in the Pamplona cases, let's see what message the referee signals to the public. Pardon my sarcasm once again. Whenever you feel frustrated as a fan, just run up into the field and kick whatever player you don't like! With a little luck, the guy responds, which earns him a red card! You obtain three results: give way to your frustration, hurt an opponent, and establish yourself as a hero among your low-IQ comrades. Are there any legal consequences? You will probably end up in prison, of which you are a regular client. What do you expect the police to do with these low-IQ marginalized people?

One may ask where exactly the legislative naivety is located in the above example. It is in the ingenuous belief that in the case of assault self-defence can be strictly proportional. Well, it is not. The weaker physically the victim, the stronger the first blow has to be, if the victim wants to make it out alive.

The naïve person would refute, "With such measures, will there not be an increase of aggression?" Of course not. An increase in defensive aggression requires a prior increase in premeditated crimes. If the delinquent knows that the victim will not be punished for a 'super-minimal' response, he or she will think twice before attacking.

A proportional minimal response is a proportional response that ends when the assailant falls: it's what the legislator thinks a normal person is able to do. A proportional super-minimal response is a proportional response that continues after the assailant falls. Thus, the fact of not punishing the super-minimal response does not increase violence in society; in fact, it will reduce it. Naturally, there will have to be a measure in that "super-minimalism". In Alvarado's case, the assailant threw a karate kick, which, if well executed, would have left him quadriplegic. That the victim ends up in the hospital, much less hurt than quadriplegic, is decidedly within the limits of super-minimalism, whence acceptable.

3.4 Abortion, suicide, homosexual unions

In this section some practical examples of Hume's guillotine will be given, which might help relieve some too heavily burdened Christian consciences. A third formulation of the 'is-ought' error is to consider it a moral duty to remedy other people's wrongdoings, especially with respect to life (abortion, suicide, euthanasia) and education (adoption by homosexuals).

If the laws of a country establish it, immoral actions are part of legal liberties, *independently of whether the majority errs.* Of course it is difficult to accept for a Christian, convinced of possessing moral truth, not to respond violently. But acting any other way is antidemocratic and even risks immorality. Moreover it is not really elegant to yell in the face of all those wretched mothers that already decided to abort (freely or under heavy pressure) or at those equally miserable butchers presenting their services, that abortion is homicide. They know it, and they hate to do it.

Whatever moral offenses occur, for a Christian it is important not to forget that from God's point of view only the perpetrators suffer, never the victims. This is explained in more depth in appendix 5. This implies that an abortion only affects mother and butcher, but it can impossibly affect the eternal bliss of the aborted. If the only fact of not having survived your zeroth birthday leads to eternal grief, what about all those natural abortions?

If God would have wanted, nothing would have stopped him from creating a world without spontaneous abortions. Well, the fact is that they do exist. Therefore, the existence of spontaneous abortions entails a good greater than its absence (plus the consequences of its absence).

3.5 Female discrimination

The cruellest male invention to continue dominating women is called 'feminism': make them believe that someday they will run 100 meters in 9.6 seconds. Truth is that the female salary is still 30% below the male's, and most top positions are still occupied by males. What's wrong with the feminist tactic?

The physical differences between men and women are written very deeply in our genes: our sexual differentiation system (XY) is even used by some insects and fish. Given that today's medical research has not even managed to eliminate the flu, it would be utterly stupid to think something can be done against the distinction of sexes. There is no other choice but to continue living with this biological distinction. As anti-feministic as it may sound, this most specifically implies that the physical distinction should be complemented with a social one.

In the society of our human and ape ancestors, the female has always been more vulnerable for her being smaller sized, and for periodically having to give birth after a long and tiring pregnancy. Moreover, the female took care of the social organization, while the male would occupy his time with naps (lions excel in it), fighting for leadership, or running after prey, in case the dumb females wouldn't manage to do the job.

All animal behaviour is chemically programmed in their genes. Species with a 'wrong' type of behaviour did not survive evolution. Again, with the poor present-day scientific comprehension of how genetic information produces behaviour, any attempt to change human gender psychology is idiotic: as idiotic as believing that the extremely complex gender behaviour can be reduced to varying the concentration of two hormones, testosterone and oestrogen.

For mammals, the role of the mother is essential during the first years of the little one, while the educational contribution of the father joins in at a later stage. When the child is about ten years old, in Homo sapiens, both parents contribute equally, though not in the same way: on average the mother's love is more 'unconditional', while the father's is more 'demanding'. Exaggeration of maternal love is called pampering, that of paternal love militarism. This behaviour is so universal that it must be written in our genes.

Remember that these genetic considerations only serve as an introduction to the gender question: to what degree and in what sense should our laws deal with gender? The second question is best answered by postulating that gender laws should strive at respecting the genetic nature of the animal in us. And as far as the first question is concerned, do not expect scientists to provide you with the answer, nor philosophers, nor theologians, nor ideologists, nor the church. The only trustworthy answer comes from nature itself. It's like a physical experiment: society has to find it out by trying several options and then choosing the optimal one.

The quality of different options should be compared using the previously introduced economic indicator. The better the laws reflect the true nature of man and woman, the stronger the economy is able to grow (all other factors being equal).

The author's ingenuous-utopian proposal is to reward maternal education by law. Simple-minds like the club of Rome would shout out immediately that such a law could never boost an economy, but who cares:

they also predicted that we would have been out of oil around 1980. For an adult man to perform economically, he must be psychologically stable. Such stability depends most on maternal care. For this reason, in case of a separation between parents, the children tend to be allotted to the mother, unless her educational inaptitude is proven (generally very difficult). Conclusion: it should not surprise that society might benefit, from a purely economic point of view, from maternal reward, if well administered.

As the author is not well enough acquainted with the British situation, I will mention two Dutch peculiarities. First, a growing number of feminists fear that the original cause of feminism has evaporated, because in spite of all of the economic and political incentives, many Dutch women still prefer to take care of their children at home over having a job outside. Second, Dutch women working outside, in academic or leadership positions, are discriminated by their male peers. It is only too easy to predict that in The Netherlands this discrimination will grow stronger and meaner, when the government continues its plans to implement hard limits of female participation in high-profile positions.

Now what is the author's utopic plan? First, in order to reward maternal educational quantitatively, a scale must be specified: all women earn

(i)   a pension proportional to the sum of taxes paid by all her children and sons-in-law;

(ii) a salary proportional to the children's performance at school; this last reward is independent of the intellectual level of the school: an excellent performance at university contributes equally to maternal salary as an excellent performance in a technical school.

The second specification (independence of IQ) helps parents making the right choice for their children. When children repeat a year at school mother misses the respective reward. The same holds for the case that children end up in prison. Women that aspire to outdoor careers are more motivated and gifted than the average. The reason is purely probabilistic, and hinges on the only fact that *human talents (independent of gender) are distributed*: the odds decrease toward excelling in all talents, and towards underperforming in all talents.

With a law for maternal salary and pension more women will choose motherhood, and such a choice reduces their financial dependence on their husbands. Consequently, the market offer of female labour decreases and due to the probabilistic law of talent distribution, the average talents of competing women on the labour market rises; whence women's salaries rise, due to both scarcity and higher average skills. In such a society, women would be appreciated much more: inside home for being irreplaceable (the little ones need unconditional love, not military demands) and outside home, for being a scarce resource with above average skills. Most importantly, such a law might reduce the appeal for young women to find a job in night clubs, and it certainly puts an end to that abject abuse which is poverty-driven prostitution. The only prostitutes left will be of the 'escort' type. This latter kind of prostitution will never disappear: not even in a society with perfect laws. Rather than a symbol of slavery, escort prostitution is a symbol of freedom, albeit a sad one.

"La donna e mobile, cual piuma al vento", complains the Duke of Mantua in Verdi's *Rigoletto*. Clearly, the Duke never experienced a female hormone cycle. The more a young woman fights against her abrupt variations of hormone levels, the more self-control she generates. When these variations reduce with time, we men will learn strength and stability of character from women. Probably a minority opinion, this one. At least the male crew of Margaret Thatcher and Neelie Kroes might agree with me.

3.6 Penitentiary reform

The British statistics indicate a rate of recidivism of 25%, and a recidivist repeats his crime on average 3 times. These numbers show there is ample room for social and above all legal improvement. First, these numbers suggest that first-time delinquents should not be imprisoned with recidivists. Why not give the first-timers a GPS ankle bracelet and let them earn money? These first-time convicts can earn money first to indemnify their victim(s), and if they have not yet done their time, they can start building up their starter's capital. Depending on the gravity of the offense, the court may rule about how many hours per week the convict should work, and about the convict's living conditions. Hume's guillotine requires that all delinquents be kept working in whatever way, (i) to maintain, obtain, or find back psychological equilibrium, and (ii) to indemnify the society that feeds them. As far as recidivists are concerned: those who are able to work get the same treatment as first-timers, along with a much longer time. Whoever is not able to work independently (drug addicts) should work under supervision. Non-violent convicts willing but unable to work may contract a loan for learning a trade. These convicts can be associated with first-timers who apply for it, in return for financial compensation.

Violent or multi-recidivist convicts will be given large palaces to live *where nobody goes in or out except the psychologist*, who may at any time decide to promote them to the GPS ankle bracelet status. This regime eliminates all physical contact with jailors, friends, or family, and thus eliminates all forms of violence. Of course, contact with family through glass and telephone is allowed.

3.7 Competence and competition

In the last section of this chapter the need is stressed to control the efficiency of government officials. The author has no specific proposal, but suggests the officials themselves devise a proposition. The idea is that one should avoid the European monster of governmental institutions and expenses, a monster the author coins 'bureaucratic socialism'. It is the opposite of the principle of subsidiarity[7], which stipulates that each issue

---

[7]   The principle of subsidiarity (the lack of which is now killing Europe) was introduced in its actual definition by a key document of Catholic social doctrine — the German precursor ("Rundschreiben über die gesellschaftliche Ordnung", §80) to the encyclical "Quadragesimo Anno" by Pius XII (1931), then Cardinal Eugenio Pacelli, papal nuncio in Nazi Germany, and author of the encyclical letter "Mit brennender Sorge" of 1937. Pacelli wrote his highly critical encyclical letter (signed by his predecessor, Pius XI, and forbidden to be divulgated or read in Nazi Germany) when the champions of naivety Neville Chamberlain (Prime Minister of the United Kingdom), Albert Lebrun (President of France), Édouard Daladier (Prime Minister of France), Hendrik Colijn (Prime Minister of the Netherlands), and Benito Mussolini (President of Italy) competed in praising their hero Adolf Hitler; *the same sillies —on mainland and on island alike— rebuked Pacelli because of his hostile attitude to Hitler.* There was only one happy exception to that parade of debility: el generalísimo Franco of Spain.
Today the great majority of the European population believes that Pius XII (the very Eugenio Pacelli of the 1937 encyclical letter) failed during WWII. The libretto for Rolf Hochhuth's play "Der Stellvertreter" (1963) depicting a silent, anti-Semitic Pope, was written and published by the Russian secret service; the same one that, a few years later, ordered Ali Agca to kill John Paul II.

should be resolved by the authority (normative, political or economic) closest to the object of the issue. The goal of bureaucratic socialism is called the "verzorgingsstaat" in Dutch, which means something like "welfare state".

Bureaucratic socialism is the guarantee of economic failure. Its naivety consists in thinking that officials can be competent without having to compete on a free market. An excessive bureaucracy has the same effect on the economy as a mafia. Where free competition lacks, nepotism and incompetence thrive.

An example of dramatic incompetence of Dutch government officials concerns the connection of Amsterdam and Rotterdam with Antwerp, Brussels, and Paris using a high speed train. The Italian offer was the cheapest (an extremely popular criterion in my country) and—oh coincidence— the Italian firm AnsaldoBreda won the international contest for the construction of the trains in 2004. Years went by. Intense litigations followed. In the year 2015, the Dutch still travel at low speed. The price of the zero km/h increase in travel speed is 11 billion Euros plus a bunch of demoralized officials.

Having government officials working without any control of their efficiency is not the smartest thing to do. Incentives can be given to officials who propose reducing the size of a department without loss of efficiency, or who propose simplifying the legal system without slowing down the economy.

Take for instance the Department of Education. It spent 57 billion pounds in 2010. That sounds like 50 billion too much. What exactly is a Department of Education responsible for? It should maintain or increase the level of education in the country. Isn't it enough to

(i)   control there is no cheating in final school/university exams;
(ii) define the matters to be learnt and write the final school/university exams ;
(iii) yearly publish the average grades of all schools/universities;
(iv) check the schools on bullying, drug consumption, and hygiene?

We presently have the very awkward situation that the Department of Education is involved both in determining the requirements presented to educational institutions, and in providing them with departmental and statutory advice. This is asking for trouble: for ever growing expenses, and for ever decreasing efficiency.

Chapter 4: Conspiracy ingenuousness

In section 4.1 some key ingredients of a conspiracy are given. The list is by no means exhaustive, nor requiring. Section 4.2 repeats some of the already well-known evidence for the conspiracy leading to the assassination of John Kennedy. It will help the reader to distinguish between terroristic acts of loners, lunatics or outsiders on one side, and those by insiders on the other. *Only terroristic acts commanded by insiders are called conspiracies.* Always keep in mind there is a huge difference between the commanding conspiracy and the executive mercenaries. Section 4.3 concerns the moon landing. Section 4.4 summarizes well-known proofs of the 9/11 conspiracy. Section 4.5 tries to connect the dots.

4.1 The assassination of Julius Caesar

A typical ancient conspiracy is the Roman one against Julius Caesar, where a few very powerful men came together to kill another. The ingredients to a conspiracy are generally the following:

(i) illegality of the action (killing without a legal verdict);
(ii) secrecy in the preparatory phase.

If the conspiracy is morally wrong, there are three more ingredients:
(iii) mutual distrust;
(iv) increase of personal power;
(iv) absence of honour and idealism amongst the conspirators;

The contemporary conspiracy includes a few more characteristics:
(v) secrecy during and after its execution;
(vi) the patsy;
(vii) invisibility of the conspirators before, during, and after the events;

(ix)   abuse of national structures (military, industrial, financial), but above all of the secret services;

(x)   careful timing of the publication of facts related to the attack;

(xi)   the production of contradictory testimonies at the moment and after the attack;

(xii) enormous blunders in planting evidence and other organizational details;

(xiii) production of conspiracy theories after the attack and physical elimination of people with crucial information;

(xiv) the immediate destruction of all evidence.

As for the ingredients (iii)-(xiv) it is possible to mention the existence of morally praiseworthy conspiracies, like that of the many German attempts to kill Hitler. The 'patsy' mentioned in ingredient (vii) is an innocent person chosen on the basis of an easily constructible motive and compromising personal circumstances and social connections. Ingredient (xii), the enormous blunders and botched jobs, is a direct consequence of the command structure. The group of conspirators consists of a few very powerful people. Due to the physical impossibility that the conspirators supervise all the jobs directly, multiple errors are inevitable. But all errors can be fixed to a certain point, and can even help to confuse simple-minded investigators.

4.2 The assassination of John F. Kennedy

The typical example of a modern conspiracy is the assassination by the CIA in 1963 in Dallas, Texas, of John F. Kennedy, American president. Wikipedia says, "A ten month investigation (…) by the Warren Commission concluded that Oswald acted alone in shooting Kennedy, and that Jack Ruby also acted alone when he killed Oswald before he could stand trail."

This is ludicrous: within ten days the patsy is assassinated (after his arrest Oswald continually claims to have been set up) and people still believe the report of the Warren Commission! Warren was either threatened or a conspirator himself.

Senator Max Leland initially was part of the Warren Commission. When he openly declared the investigation to be compromised[8], he was replaced. Here, Cleland accuses the government of delaying or denying access to vital documents; note that his order was to report to Congress in May 27, 2004. In November 23, 2003, UPI announced that, "the past Senator Max Cleland has been nominated to be a member of the Export-Import Band by President Bush. Therefore, he will have to abdicate the Commission that investigates the terrorist attacks of 11-S." He was replaced by Senator Bob Kerrey, a war criminal and fan of PNAC (Project for the New American Century, a neoconservative think tank founded by Dick Cheney in 1997).

For whomever wants a definitive proof of the fact that there were six shots (none of them fired by Oswald) it suffices to refer to Randolph Robertson's study, in which he synchronized the video footage by Zapruder with the acoustic tape "DictaBelt" from the motorcycle policeman, H. B. McLain. This synchronization, made with a precision of a few milliseconds, not only counts six shots and their multiple echoes, but it also identifies the three sites where the shots originated from. Four of the shots reached the president and a fifth wounded Governor Connally.

Just before the assassination, the motorcycle guards had been ordered to retreat. The Warren Commission established that the protocol anomaly was ordered by Winston G. Lawson, CIA. Obviously, when these three letters appear in that sequence, the investigation stops. The logical question for Lawson would have been to know from whom he received the orders, but the Commission was more interested in Lawson's personal feelings at the time ("I thought the president did not appreciate motorized policemen around").

Wikipedia mentions that in his death bed, CIA Howard Hunt accused Lyndon B. Johnson of being responsible for Kennedy's assassination. Well thank you very much for this confession, Mr. Hunt, but you are late and what we need are proofs, not accusations. There were others, more

---

[8] "If this decision stands, I, as a member of the commission, cannot look any American in the eye, especially family members of victims, and say the commission had full access. This investigation is now compromised", see http://www.oilempire.us/investigation.html

powerful than Johnson and who benefited less patently from Kennedy's death. It is easy to imagine the CIA's frustration after their humiliation in the Bay of Pigs in Cuba (1961), followed by the missile crisis in 1962. In 1963, Kennedy signed the National Security Action Memorandum 263, which ordered withdrawing 1,000 military from Vietnam. It is very probable that Kennedy had in mind to withdraw the rest of the army from Vietnam upon re-election (in 1964). His speech in June of 1963 given before the American University of Washington D.C. was entitled "a strategy for peace". Notably, he commented that the United States were trying to approach the Soviet Union in order to begin a bilateral nuclear disarmament, and that the U.S. would never start a war against the Soviets. Imagine the Zionist panic! After his Washington speech Kennedy already was a walking dead man.

4.3 Lunar expedition

It is interesting to think about the large amount of people that think that the lunar expedition never happened. A convincing proof that it did happen is that the then only competitors, the Russians, (who had more powerful equipment than the critics, to follow the American operations in lunar space and to verify their claims) never challenged the US the moon landing. Whoever believes the lunar expedition to be a conspiracy is a crackpot.

4.4 New York September 11, 2001

The craziest conspiracy theory the author ever heard is described in the official 9/11 report. According to that report, nineteen Islamic fanatics seized command of four planes at exactly the same time, confusing the air traffic controllers during several essential minutes. The controllers thought it was a CIA exercise, because there was one planned that very day in California. The internationally admired American defence system was not able to get a single fighter jet at even 50 km from the four seized Boeings. Although the Russians and Chinese have seen all the details, American civil radar, military radar, and satellite radar either saw nothing or trashed the relevant information. The whole Pentagon was being watched by only one surveillance camera. A Boeing manages to disappear into a tiny 16 feet (5 meter) diameter hole in the Pentagon. The vast operation was being coordinated from a dark Afghan cave by the twentieth conspirator, a CIA-resurrected version of Bin Laden, who was probably three years gone. After a masterly pirouette two musketeers crash their Boeings into the Twin Towers. Even more masterly, the second musketeer manages to make everyone believe his Boeing looked like a military drone KC767! In the meantime cell phone calls come in from a fourth plane at cruising height, even though there is no coverage at that altitude. Firefighters are able to control the twin tower fires in not more than a quarter of an hour. Molten steel is seen flowing out of one tower, revealing twice the temperature (on a Celsius scale) an ordinary combustion fire can ever reach. Physical miracles abound. An Islamic passport survives the molten steel, which smells an awful lot like Charlie Hebdo in Paris, and Madrid/Vallecas in Spain. At the moment of impact the president is assisting to a class of schoolchildren learning how to pronounce the words "airplane" and "steel" (imagine the laughter of the CIA members who suggested these words to the teacher). Then the president is discretely informed "that the nation is under attack". Instead of being thrown to the ground by three bodyguards and taken to the nearest safe house, the President sits still and smiles.

Wikipedia discloses the testimony publicly given by Norman Mineta, secretary of civil transportation, who ordered the suspension of all domestic flights. This testimony, requested by the 9/11 Commission, was not included in the final 9/11 Commission Report. Mineta was in the White House bunker and testifies on the moment flight 77 of American Airlines was headed for the Pentagon: "There was a young man who had come in and said to the vice president, 'The plane is 50 miles out. The plane is 30 miles out.' And when it got down to, 'The plane is 10 miles out,' the young man also said to the vice president, 'Do the orders still stand?' And the vice president turned and whipped his neck around and said, 'Of course the orders still stand. Have you heard anything to the contrary?' Well, at the time I didn't know what all that meant."

The real meaning of this mysterious conversation became clear to Mineta after the disaster; *the vice president was referring to his own orders that all fighter jets should be kept on the ground! This can only be understood if the impact of the missile on the Pentagon was part of the plan.* Convincing the civil American population of the danger of Islam was simple enough (just show them some compatriots jumping from the Twin Towers), but for the military hierarchy made up of tougher men something more was necessary: an attack on the Pentagon.

The explosion at the Pentagon was not caused by a Boeing but by an ordinary missile (//watch?v=5cFewUG3rSY). The images that American civilians continue to upload on Youtube (and the conspirators continue to destroy) show the impact of a missile in a rather convincing way. In the debris there are no signs of any Boeing parts or of human bodies.

The crash into the Pentagon had a farther-reaching goal. A day before the attack, September 10th, the secretary of defence, Donal Rumsfeld, made an astonishing declaration, reported on January 29, 2002: "According to some estimates, the destination of 2,300 billion dollars cannot be located." PBS (Public Broadcasting Services) revealed that this figure comes from the very inspector general of the Pentagon. It is of the order of the Gross Domestic Product of the whole of the United Kingdom! The news that Rumsfeld gave, was not a mere explanation of a list of expenses, but a shameless admission that the military do not have any intention of reporting to society. It is clear that in ordinary circumstances Rumsfeld would have been shredded alive in Congress. But what a coincidence, the day after happened to be 9/11 and the 2300 billion dollars were no more spoken of.

Some say that the 2,300 billion dollars should not be taken literally, because they are the consequence of a system that is very old and is fragmented in its communication of facts. If it were so, why did the secretary of defence, Rumsfeld, come up with another cause? In addition, there will always be a central computer that distributes the received sum of the government in small parts. Did this central computer also break down? Well, in this case, I suggest doubling the military budget. Maybe with 4,600 billion dollars the military can buy a computer to register its expenses.

For the interested, here follow some other issues with the official 9/11 report.

First, the Pentagon was hit by an object traveling at 900 km/h. No Boeing 757 can exceed 600 km/h at ground level because the air resistance is too high; if it were forced to go 900 km/h per hour (through added motors) wings and fuselage would succumb to air turbulence. Hence the object was a simple missile. This agrees with the sound it made according to the witnesses, and with the diameter of the hole in the concrete wall (16 feet or 4m90). On a Spanish website[9] Dave von Kleist proves that such a hole could never have encompassed a Boeing 757; that the damage is totally incompatible with the amount of kerosene that must have burnt. Wooden tables and chairs are clearly visible, even books laying open with undamaged pages that survived the impact.

Second, the supposed flight UA175 that crashed into the second WTC tower was not a Boeing 767 (twin engine, big grey/black plane, no marking, no emblems, no logos, no windows on the sides, witnesses say), but very similar to the smaller military "drone" KC 767 with a modified belly and attached missile. When a plane crashes into a concrete tower with windows, there has to be some percentage of the plane that bounces off and falls to the ground. However, the images show that the tower sucks up the entire plane. In almost all images one can even see the explosion of the missile fired by the drone just before impact. This increases the temperature locally above the nanothermite ignition temperature, creating a large hole in which the complete drone can disappear with nothing bouncing off.

Third, on the Shanksville tapes, at the end of the conversation, a sigh of UA93 Flight Attendent CeeCee Lyles can be heard, whispering, "it's a frame". Probably she had that courage because she knew she would be executed anyway.

Many videos shown on Youtube have been visibly manipulated. In one of them a plane is seen passing behind a crane, but it in front of its cables (/watch?v=IL8cJWyOxWQ). In another, an explosion is clearly seen taking place in the building before the impact of the plane (/watch?v=mhROd7Jt3-w). Most of the impact tapes show that the drone fully enters the building, instead having its wings bounced off. One video shows that the so-called Boeings hardly made any sound — the public is seen reacting on the sound only at the impact, the noise of which most probably was due to the drone's fired missile (look at / watch?v=CUoqwUVOxHE). This video also mentions that the military radar did not detect a commercial flight at the WTC, but 300 meters off.

---

9   http://www.bibliotecapleyades.net/sociopolitica/esp_sociopol_911_90.htm

The passengers of both flights (AA11 and UA175) supposedly crashed against the towers WTC1 and 2 had the same treatment as those from UA77. With an added detail, that their mortal remains were scattered all over ground zero, so that forensic medics could pick up some of their DNA. The CIA bunglers fulfilled criterion (xii) of the contemporary conspiracy by planting the wrong motor on Murray Street: they planted a CF(M5)6 from General Electric instead of a Pratt&Whitney used by Boeing. Fighter Jet pilot John Lear sarcastically concludes in his interview, "the guy dropped off the wrong engine."

How could they have disappeared without leaving a trace of the four hijacked civil flights: UA77, AA11, UA175, and UA93? It was not out of pure coincidence that the CIA planned, that same day 9/11, a secret operation of aero hijacking in California—obviously to assure that all the useful and competent personnel in aero hijacking would have been far from the command posts.

History Commons Alert describes the schedule of the exercises of the North American Aerospace Defence Command (NORAD). At 9:40am, on September 11, the "hijacking simulation" would begin, predicted as part of the NORAD exercise. Upon hearing the report of the hijacking of flight AA1, Kevin Nasypany, the NORAD commander of NEADS, Northeast Air Defence Sector, member of the team that put together the exercise, deduced that "someone had started the exercises very early." Not by incompetence. Everything was predisposed to sow doubts.

The flights probably landed in Westover Air Reserve Base (Massachusetts), owing to its extended landing strip necessary to land planes heavily laden with fuel. The calls were made from there, in order to convey the message that the hijackers were Islamists.

Fourth, the sequence of debris planting in Shanksville is an ode to criterion (xii) of contemporary conspiracy. The official 9/11 report declares that flight UA93 crashed in Shanksville, Pennsylvania, but all of the witnesses emphatically deny it. "This crash was different: there were no debris from the plane, no bodies, and nobody heard anything" (Wallace Miller, forensic doctor of Somerset County); "I was looking for anything that said tail, wing, plane, metal; there was nothing" (Scott Spangler, photographer); "I was amazed because it did not, in any way, shape, or form, look like a plane crash." (Patrick Madigan, police commander of

Somerset barracks of the Pennsylvania State Police). "Although flight UA93 was reportedly 'heavily laden with jet fuel' when it crashed, investigators found no contamination of jet fuel in the soil and ground water around the site"; and "The Independent reported that flight UA93 was carrying 3,410 kilograms (7,500 pounds) of mail to California when it crashed." Like in the Pentagon: again no marks of combustion at all. The above quotes can be found on http://911blogger.com/news/2013-02-19.

FBI agent, Wells Morrison comments, "It was interesting. The black boxes are next to each inside of the plane, but they were found at different depths, separated by 4 meters (13 feet)." Morrison also commented that he was "truly surprised that the boxes were not found earlier", which is a polite way to say that the CIA bunglers were much too late.

Fifth, according to the official conspiracy theory there were twenty terrorists, nineteen of which on American soil. In those days, after the incident, the list of names would change almost daily — as soon as an assumed terrorist produced an alibi, another name had to be found. From the final list, there are five that are alive today. Another formidable botched job.

Sixth, the public has largely forgotten that three towers came down on 9/11, not just two. The third building (WTC7), also known as the Solomon Brothers Building, collapsed, too. This giant of 47 stories contained among others the national archives of financial scandals (like the scandal revealed in 2001 of the American company, Enron). It came down in seven seconds, without any preceding damage, clearly the result of a controlled demolition. The official 9/11 report again denies what is evident. The occupants of the WTC7 building on 9/11 included the Securities and Exchange Commission, the Internal Revenue Service (IRS) Regional Council, the Department of Defence (DOD), and the Central Intelligence Agency (CIA). Interestingly, there was nobody in WTC7 on 9/11. Forty-seven empty floors on a normal working day! How is this possible without previous warning? Indeed it is not, and not only all the personnel working in WTC7 knew it would be blown, but also the New York fire fighters, and even four international broadcasting companies! Dr. Graeme MacQueen's in-depth analysis is worth seeing, as well as all the lies and misinformation of BBC trying to hide their multiple blunders with Jane Standley. She mentioned live and on-site the collapse of WTC7 while it was still standing, right behind her.

Seventh, during an intensive renovation of the Pentagon, all civil workers were sent to Washington D.C. Only 65 of them (all employees from Resources Services) returned to the Pentagon after the renovation was completed: mainly civil accountants, analysts, and budget managers. They had professional knowledge of the destination of 2.3 trillion dollars. The official 9/11 report declares that, "due to the impact of Flight UA77, unfortunately, 34 civil servants also died." And with them disappeared the crucial information of the origin and destination of 2.3 trillion dollar.

Eighth, physico-chemical analyses of the thick (10 cm) layer of dust that covered the whole city after the 9/11 attack were published in *The Open Chemical Physics Journal*, 2009, **2**, 7-31, by nine scientists among whom Steven Jones in a scientific article entitled "Active Thermitic Material Discovered in the Dust from the 9/11 World Trade Center Catastrophe". It ascribes the collapse of the twin towers to a largely unknown explosive called "nanothermite." The authors were able to reconstruct the exact shape and composition of the explosive material. Years before its use the military properties of nanothermite were extensively described in a National Laboratory Journal. It was shown that nanothermite cuts through steel like butter, and pulverizes concrete walls like sand castles on the beach. Its advantages over classical explosives are mentioned: ignition occurs at a single critical temperature, not along a trajectory; sudden large pressure changes do not set off the explosive; the explosive is applicable in the form of wall paint.

Just like in the Spanish attack in Madrid (Vallecas) all evidence was deliberately destroyed (characteristic xiv of the conspiracy). In this case it implied the enormous job to transport all evidence of cut steel (that is, *all steel of the two twin towers*) to India, and have it molten asap.

Remind that the Mineta testimony was not published in the final report, but the military properties of nanothermite were published: selective publication is characteristic (x) of a conspiracy. This way no smart-ass investigator could claim to have found a secretly developed weapon. There are only three places in the world where the necessary amounts of nanothermite could have been produced for pulverizing the 440 meters high twin towers: Los Alamos National Laboratory, Lawrence Livermore

National Laboratory, and a chemical laboratory of the Naval Surface Warfare Center. Ever since the discovery, the practical development and design of the production of these large quantities need at least a decade, probably two.

4.5 New York, London, Madrid, Paris: what connects the dots?

Northern America is not the land of unlimited opportunities anymore. This year, for the first time in history, Mexican return surpassed emigration to the US. Let's have a look at some basic facts, generally known:

- Americans have the most advanced technology in the world in a lot of industrial sectors;
- they work significantly more hours than Europeans;
- they have a relatively slender bureaucracy (i.e. no waste of money on incompetent officials);
- Yet, they have a deplorable social security system and severe poverty.

In one sentence: even though Americans produce more and spend less, they are visibly poorer than the Europeans. This is only possible if their parasite is significantly bigger than ours. We all know what the parasite does in the U.S. It steals 2.3 trillion dollars for playing war in the Middle East. But what exactly are the revenues, apart from a continuous stream of coffins? Would any American, healthy of mind, choose to embark in endless fights over nothing but a few oil mines, knowing that oil will be obsolete in a few decades from now? The author's guess is not: just as in the final years of Kennedy, most American citizens are fed up with wars that seem to be nobody's cause.

Nobody's cause? Actually, there is a very small group of people very interested in these wars: Zionists, a very small minority of faithless Jews who strive at dominating the world. Destabilizing the Middle East is just the first step in their strategy. If Muslims keep fighting among each other and everybody else, it is only a question of time and Israeli tanks will roll in and "pacify" these war-stricken regions: Syria, Iraq, and possibly Iran and Jordan. Of course, the naïve western Christians will witness such pacification with awe, applauding the Zionists' ability to end the Muslim wars in so little time.

The clue to the Zionists' involvement is the June 1967 fate of the USS Liberty. As is explained in detail[10] the attack was premeditated and conscientious, and deliberated, killing 34 sailors and injuring 172. According to James Ennes, Liberty's personnel received firm orders not to say anything to anybody about the attack, and the naval inquiry was nothing but a cover-up. Admiral Tom Moorer stated that the Liberty was the most identifiable ship in the U.S. Navy and in an interview with the Washington Post stated: "To suggest that they [the Israeli's] couldn't identify the ship is ... ridiculous. Anybody who could not identify the Liberty could not tell the difference between the White House and the Washington Monument."

One might think an enormous diplomatic crisis would follow: nothing of the kind. The sailor's families received a ridiculous compensation. Children under five were not even counted "because they would have forgotten their father". The way the Americans handled this incident is vomiting: a humiliating genuflexion to Zionism. Only one conclusion is possible here: the Zionists already had enough of the American Congress in their pockets, obviously by paying all their expensive election campaigns, and making sure some compromising photos were available in case of incompliance. From the 9/11 experience and Kennedy's assassination one may deduce the Zionists also control enough of the CIA. The CIA had no issue with a president that wanted to end the middle-eastern wars, but the Zionists had.

It is a sad truth for us Europeans to conclude that our own secret services are also fully controlled by Zionism. Who profits from all the Syrian refugees, from the European bombings? Not the Christians, nor the Muslims, nor the Americans, nor the Jews. But the Zionists do profit: when Christians and Muslims fight each other, they reign. To state it explicitly: all of this (New York, Spain, France, and London) was ordered by the Zionist secret services, and obediently executed by the local secret services.

The Spanish secret services executed Mossad's wishes superbly. Ordinary Spanish people have no idea what happened on September 11th,

---

10 https://en.wikipedia.org/wiki/USS_Liberty_incident#Ongoing_controversy_and_unresolved_questions

2004. They have all been fooled by the timing of the attack: just before the elections. Everybody thinks the Madrid attacks have something to do with the elections. In the Spanish version of this book it is explained in detail that all conspiracy characteristics were present, including the spectacular destruction of all thirteen remaining evidence samples, and the sudden appearance, on site, of a passport identifying the terrorist. In Spain too, the official report is a collection of inconsistencies. It is not even known whether people really died in Vallecas or they were just agents with moulage (make-up) like in London.

An equally nice job was done by the MI5 in the 2005 tube bombings on July 7th in London. The Shadow Home Secretary, David Davis, finally understood it: "It is becoming more and more clear that the story presented to the public and Parliament is at odds with the facts." Anything more explicit would have jeopardized his life (or that of his family). Nobody was injured that day. The most notable example is the famous "white-masked woman".[11] Davinia Turrell (now known as Mrs. Douglas) appears on all the photos with both hands intact although she declared herself: "There was a loud bang and a ball of fire appeared from my left hand side and seemed to go right round me and then quickly retracted. After the explosion, the carriage was actually very quiet. Everyone was too shocked to scream or shout." Oh yes, the perpetrator's name is Mohammad Sidique Khan. MI5 counts on it that the British average IQ is high enough to recognize this as an Islamic name.

The French secret services executed the Zionists' wishes equally well. "Speaking to PressTV, Jack Linblad, of the Green Party of Los Angeles County Council, claimed that the actions of the Kouachi brothers and Amedy Coulibaby, who collectively shot and killed 17 people in three separate and shocking attacks in Paris, were not acting on their extremist religious beliefs but were instead carrying out orders from the US and Mossad. Mr Linblad added that he believed the bloody attacks had been

---

[11]   http://nodisinfo.com/white-masked-woman-london-tube-bomings-hoax/

orchestrated to keep Europe under Netanyahu's thumb and to ensure the Israeli leader stays in power."[12]

For those who still doubt the French attacks are an inside job: a passport of the perpetrator was found on the scene, just like in Madrid and on ground zero. Seek, and ye shall find... all fourteen conspiracy ingredients.

The second step in the Zionist strategy is all too easy to guess: to "pacify" Syria, Iraq, Jordan, and Iran using combined NATO-Israeli military equipment, after unanimous NATO approval, under crashing Christian applause.

12 http://www.independent.co.uk/news/world/europe/paris-attacks-us-politician-jack-lindblad-claims-charlie-hebdo-killings-were-by-us-and-mossad-to-9979696.html

Chapter 5: Ideological ingenuousness

The road traffic in Dutch Randstad (Amsterdam, Rotterdam, Utrecht, Leiden, Delft, Den Haag) has grown enormously in the last century. For many decades, there has not been any project of proportionate growth in the highway network. Five years ago, our transport officials had converted the Dutch freeways into a permanent traffic jam—bigger than the one in Madrid, Los Angeles, and New York. The officials with greatest authority tried to explain to the Dutch people that expanding the highway network was impossible due to ecological and structural reasons. Three years ago, the Minister of Infrastructure and the Environment, Melanie Schultz van Haegen-Maas Geesteranus, belonged to a right wing party. Not bothered by ideology, she started to build roads. What happened? The traffic jams decreased. Clearly, ideology is a 20th century relic that is able only to cause economic catastrophes.

For every country, the possession of laws encouraging economic growth, and people determined enough to fight corruption, is a blessing. Look at history: communism, fascism, ecologism: ideologies only screw things up; they never contribute to social wealth. In the first section of this chapter the pernicious consequences for society due to educative ideologies are discussed (5.1). Afterward Al Gore's ecological Neo-Manichaeism will be reviewed (5.2). The chapter ends with the danger of educative ideologies (5.3), and suggests a way to de-ideologize education.

5.1 Ideology in education

When the English society attacks a Catholic school for not teaching the equality of heterosexual and homosexual union, it would be a nasty error to appeal to the liberty of education or of religion, because any success along these lines will be abused by Muslims to introduce their immoral sharia.

The Koran was compiled in the 7th century, and looks like a law book of a Christian sect, with copious mention of Christ and his Mother. It contains 114 suras (chapters) categorized by their length (number of verses, or 'ayah'). Apparently that is the pedagogical method God chose to inspire his prophet. Approximately a third of the Koran is eschatological. The rest is source of sharia (Islamic law), together with the 'Hadith' (jurisprudence of the 8th and 9th centuries). Obviously, these ancient laws are retrograde when seen through modern eyes. Hence, the only way for Muslims to live according to the sharia is to appeal to the liberty of religion. The worst problems of the sharia are its anachronistic hostility toward non-Muslims, and its strong discrimination of women.

Instead of complaining about the legal acceptance of homosexuality in England it would be advisable to teach the students from secondary school to appreciate the following nuances:

- Biologically speaking homosexual acts are as natural as heterosexual acts.
- Spiritually speaking the differences between heterosex and homosex are enormous. Statistics ruthlessly indicate (i) that homosexuals suffer more psychologically than heterosexuals; (ii) that homosexual unions are less enduring than heterosexual marriages; (iii) that kids do not want to be special for fear of being bullied; (iv) that having two fathers or two mothers increases the risk of being bullied. It pertains to democracy, not to the churches, to establish the convenience of homosexual couples adopting kids. If a society were to decide this, it would not make sense to take up arms. Remember section 3.4: the only thing that does make sense is to produce scientifically sound statistics. Believe me, the most vehement opponents of adoption by homosexuals will be the very adopted children themselves: so give them the time to grow up and accuse their parents.
- According to the Catholic religion and many Christian ones, (i) all sexual acts outside (and some even inside) matrimony are sinful; (ii) some are called 'abhorrent' in the Old Testament (e.g., homosex or sex with animals); (iii) God loves homosexuals as much as he loves heterosexuals; and (iv) God wants homosexuals to abstain from all sexual activity, just as priest do.

This is nothing but yet another example of Hume's guillotine applied to a Christian error: understanding 'you should sin, by preaching the morally illicit' when the government only wants you to explain homosexuality in your schools. Explaining homosexuality without statistics is an ideological fight, which degrades both parties.

5.2 Ecological Neo-Manichaeism

The old Manichaeism considered matter to be bad and spirit good. Today, Al Gore resuscitated a variation of this Manichaeism, which is here called 'Ecological Neo-Manichaeism'. It considers virginal and pristine nature as good, and human intervention bad. Typical of Ecological Neo-Manichaeism is certain hostility toward the globalization of the market, finances, and technology; these terms tend to be accompanied by epithets like 'blind', 'schizophrenic', or 'power abuse'. Market, finances, and technology are just tools, like a knife which can be used to cut meat or bread. The knife is not discredited for being wrongly used by a thief. Such contempt proceeds from the 'is-ought' error: assigning a moral value to a non-moral entity. The knife is not morally good or bad, but the person that uses it. The natural good of the knife is its being sharp. Ideological Neo-Manichaeism has been useful for demanding birth control programs in Latin-American countries, because the North Americans, who advocate this ideology (a very small minority, but with a very large influence in U.S. authorities that decide where the funds for poor countries will go) do not want the Latin-Americans to become competitors.

In his encyclical Letter "Laudato sí" the Pope tackles, in the tradition of the social doctrine of the Catholic Church, a wide panorama of anthropological issues that are connected by the respect for life and nature. Most ecologists are good-hearted, virtuous citizens that are worth much more than an encyclical Letter to be won for our faith. However, in doing so, one must tread carefully, in order to avoid supporting Al Gore's ecological neo-Manichaeism.

Take the typical Gorian sentence that "the earth is becoming more and more a huge deposit of junk". In practice, it is difficult to quantify the pollution caused by humans. One of the few serious efforts to quantify it was carried out thirteen years ago by Bjørn Lomborg, who compiled his conclusions in his classic work "The Skeptical Environmentalist" (2001). He points out that the cleanest countries are the richest ones. The reason is simple. Whoever opens the window in the morning does not want to see a street full of trash, strongly smelling of garbage; rather, one prefers to see a clean street smelling of flowers, if possible.

Anecdote: a friend of mine lived for some time in Singapore. After three days, the local police paid him a visit to inform him about Singapore cleaning customs; what things go into what can. They even had proofs of his misbehaviour, in the form of photos kindly provided by his neighbour. Of course, we are not talking about throwing trash in the street, but of putting a trash can on the wrong spot. One can be sure that there is no need to motivate people in Singapore to take care of the ecology. There, ecology risks being turned into the pseudo-religion of Mother Earth or "Gaia," as it is known in "New Age" spirituality.

In poor countries the situation is totally different. Naturally, a poor population has the same desires as a rich one, but the funds are lacking. In these countries it is a distinctly bad idea to concern oneself with ecology, because it would further hinder an already weak economy. Essential is that these countries acquire wealth. Once wealthy, ecology comes spontaneously. And to accumulate wealth, everybody knows the recipe: (i) increase the density of the population, (ii) eradicate corruption, and (iii) use the tools of a controlled free market, finance and technology to their full potential. As the world population grows, the rich countries grow accordingly. The Gorian statement that "The earth, our home, is beginning to look more and more like an immense pile of filth" is distinctly misleading.

That rich countries export their trash to poor countries has nothing to do with the impossibility of transforming that trash into valuable material. If the buying country is so corrupt that all of the money destined to the conversion of this trash ends up in the pocket of some patriot-playing military, the trash seller is not to blame for it.

Lomborg points out in his cited work that the prices of raw materials reduce structurally, in spite of a growing world population and the concomitant increase of consumption. All of this is possible thanks to a well-developed and efficient recycling technology. The Gorian statement that "our industrial system, at the end of its cycle of production and consumption, has not developed the capacity to absorb and reuse waste and by-products" only holds for the poorest countries.

The sea damage due to the Prestige that sunk on November 13, 2002, 250 kilometres off the Galician coast, was declared overcome on October

8, 2003, and all of the fishing restrictions were lifted. The microbes ate 77.000 tons of petroleum without even farting. This self-cleaning power of planet earth is called 'bioremediation'. According to the EPA, bioremediation is a "treatment that uses naturally occurring organisms to break down hazardous substances into less toxic or non-toxic substances."

Ideologists easily succumb to the temptation of falsifying data. A nice example is given by "The Hockey Stick Illusion: Climategate and the Corruption of Science" by Andrew Montford (2010). Montford's book describes the first failed attempts by Steve McIntyre to replicate Michael Mann's Hockey Stick graph (so called because of its exponential growth of the temperature of our planet). So McIntyre asked himself where Michael Mann obtained his data from. An investigation made by the congress concluded that Mann's Hockey Stick data were made up. Montford's book also describes in great detail the emails sent from the Climatic Research Unit. Its publication forced the director, Phil Jones, to resign. The immorality of "climate scientists" world-wide had been laid bare. Why would these scientists act like this? Well, not out of pure evil, but out of pure naivety. They have an obsession with the excessive population of our planet and they use global warming as an excuse.

Another Gorian statement refers to the "very solid scientific consensus which indicates that we are presently witnessing a disturbing warming of the climatic system". Such a consensus does not exist at all.[13]

5.3 Ideology in British Parliament

On December 4th, 2015, Reuters reports the following:

> "Britain joined U.S.-led air strikes against Islamic State in Syria on Thursday, but Vladimir Putin issued bitter new denunciations of Turkey for shooting down a Russian plane, demonstrating the limits to international solidarity. British Tornado jets took off from the Royal Air Force base at

---

[13] Elmer Beauregard writes in Wikipedia that Gore's prediction that 2014 would be the hottest year was falsified: his arguments are quite convincing (although they do not yet stand scientific scrutiny) and can be found on http://www.globalclimatescam.com/opinion/top-ten-reasons-climate-change-is-a-hoax/.

Akrotiri in Cyprus before dawn, hours after parliament in London voted 397-223 to support Prime Minister David Cameron's plan to extend air strikes from Iraq to Syria. Britain said they struck oil fields used to fund Islamic State. There are plenty more of these targets throughout eastern, northern Syria which we hope to be striking in the next few days and weeks,' Defence Secretary Michael Fallon said. Britain would send eight more warplanes to Cyprus to join the missions. The British contribution forms only a tiny part of U.S.-led 'Operation Inherent Resolve', which has been bombing Islamic State in Iraq and Syria for more than a year with hundreds of aircraft. Previously, the small British contingent participated in strikes on Iraq but not Syria."

This is a standard example of Zionist procedures: first create Islamic fear among an occidental population, then abuse of the American army to realize their goals.

What exactly was debated in the British parliament? Does anybody there know whether Islamic State is funded by the U.S.? Then why strike oil fields? What does "There are plenty of these targets" mean? Does it mean, as usual, "the U.S. obtains the coordinates from the Zionists, we get them from the U.S., and we have no fucking idea what we are bombing"? This sounds very Dutch to me and it makes me sick.[14]

---

[14] O yes, we Dutch had our own baptism: July 11th 1995 in Srebrenica. Sure, the French responsible Lieutenant-General Bernard Janvier refused to send any help. But probably he did so in agreement with the Dutch Military commander-in –chief (General Hans Couzy) and minister of defence (Joris Voorhoeve), who actively sabotaged air strikes on the Serbs, in order not to endanger the lives of 30 Dutch military fettered by the Serbs. Deplorably so, the United Nations had sent a much too lightly armed battalion to Srebrenica, communicating on forehand to the Serbs that no war would be waged. No single act of heroism among the 900 Dutch soldiers is known to the author. On the other hand shameful photos of partying soldiers abound. What did the Dutch political and military top decide? Promote chief Thom Karremans, and provide all soldiers with an honorary badge (https://nl.wikipedia.org/wiki/Val_van_Srebrenica). Ingenuousness generates cowardice generates utter stupidity.

Does anybody in the British Parliament understand why Russia does not retaliate when the Turkish shoot down a Russian bomber jet? Wasn't the Russian bombing meant to destroy Islamic-State camps in southern Turkey?

Sure, but obviously the Americans and Russians, allegedly fighting against the Syrian rebels, have quite different agenda's.

What good did it bring to Europe by fighting and replacing all those mean leaders: Mohammad Reza Shah Pahlavi of Iran, Muhammad Ghaddafi of Libya, Saddam Hussein of Iraq, and now Bashar-al-Assad of Syria? We are turning the lives of millions of innocent Muslims into hell, and creating immigration currents able to destabilize a complete continent. Did any of this come up in the British parliamentary debate? Or was it just a petty show of party discipline? Isn't it a bit strange that, on a topic of foreign policy, there is not a single MP with an own opinion, which happens not to coincide with the party's official stand?

5.4 Ideology and family structure

This last section formulates an ideological law (cfr. appendix A5). Since no proofs exist for ideological positions, the author proposes a quantitative comparison. Among the Dutch inhabitants of Moroccan and Antillean descent, crime probability is three times that for natives of the same socio-economic level. The rate of unemployment for these groups presents the same factor of three. In the United States, Afro-Americans play this role. They are imprisoned approximately three times more than other Americans (this is weighted by the difference in average social level). The official sociological studies, either Dutch or American, indicate that the primary causes are poverty and racism. Regrettably, such analyses display a hypocritical lack of imagination: poverty and racism are just as much the consequence of social problems, as their cause.

What relevant correlations do these reports come up with? None that the author knows about. A clear correlation exists however between crime and quality of family structure. Why is this never mentioned in official reports on poverty and racism? The reason is clear: it is ideologically not done.

In African families in the United States, the father tends to be absent. When the American democrat, Daniel P. Moynihan, published his famous study "the Negro family" in 1965 foreseeing a disaster, the rate of extramarital births was 25% among Americans of African descent. Unfortunately he predicted it right: in 2011 the rate of extramarital births rose to 72%, and it continues to do so.

Dutch families of Moroccan descent often lack maternal authority. But of course, the mother is not physically absent. The home is her prison, the director of the prison is her husband, and the warders are her male children. Why don't the official studies on Moroccan crime mention the discrimination of women in Islam? It is the primary and sufficient cause of a damaged family structure, which on its turn is the cause of the social problem of the Moroccan boys. Again, ideology stands in the way.

The statistics do not lie, and they allow to establishing a spiritual law: without the education of a father and a mother, united in a faithful marriage, with strictly equal authority and responsibility, the psychological stability of the children is in jeopardy.

Traditional Chrisitians mention that Saint Paul wrote that the man is the head of the woman and that she owes him obedience. As for Saint Paul's opinion, it is clear that he defended the strict spiritual equality of men and women; indeed, Saint Paul also wrote, in his letter to the Ephesians (5:25), "Husbands, love your wives, even as much as Christ loved the church." In other words, the husband's love should be stronger than the torture inflicted on Christ. Such love is unthinkable without strict equality of man and woman. In the same way, St. Paul asserted the fundamental equality between a slave and freeborn before God, leaving all responsibility to end that discrimination to society. In his declaration about the fundamental equality of man and woman, St. Paul agrees with the solemn declaration from Genesis (1:27): "God created man in his image; in the divine image he created him; male and female he created them." This declaration is not compatible with a practical interpretation of 1 Cor.11:3 (the male "directs" the female). In the same context of 1 Cor.11, it is clear that St. Paul develops a theological argument. Indeed, the first Letter to the Corinthians (11:8-10) continues saying, "For man did not come from woman, but woman from man; nor was man created for woman, but woman for man; for this reason a woman should have a sign of authority on her head, because of the angels." In this case it is advisable to leave theology to the theologians.

I will allow myself to mention one theological banality on the creation of man as an image of God, because it nicely fits in with the main point of the book. The resemblance to God necessarily refers to the spirit, not the body, because God is a purely spiritual entity. The incarnation of Christ has nothing to do with "the image of God" from Genesis, since incarnation is neither necessary for divine existence, nor for human existence. Theologically, Jesus humiliated himself by taking the form of man; not the other way round.

And since we are already talking about banalities, here are a few biological ones. The male chromosomes contain all the genetic information of the species, thereby making an exclusively feminine humanity genetically impossible, no matter what feminists say; neither is it genetically possible to have a virginal birth of a male child, because of the absence of sexual Y chromosomes in women. Thus, from all miracles related in the Scriptures, the virginal birth of Jesus is the most spectacular one from a biological point of view.

This is a beautiful example of divine exquisiteness: God allowed both sexes to contribute to salvation by incarnating male after prior consent of a virgin.

Appendices

A1: God and the world
A1.0 Summary
A1.1 A short history of the manual
A1.2 Consensus or necessity?
A1.3 Objectivity and dogma
A1.4 Philosophical manuals
A1.5 Science and ideology
A1.6 Certainty and universality of scientific knowledge
A2: The human spirit
A2.0 Summary
A2.1 Mathematics: Kurt Gödel's Incompleteness Theorem
A2.2 Biology: Intentionality
A2.3 Anthropology: Subjective claim of inalienable rights
A2.4 Economy: Lomborg's Law
A3: Quantum mechanics
A3.1 Measurement and prediction
A3.2 The Copenhagen interpretation
A3.3 Correlation
A3.4 Bell experiments
A4: Causality
A4.1 Classical hylomorphism
A4.2 Quantum hylomorphism
A4.3 Quantum causality
A4.4 Philosophical causality
A5: Laws of the Spirit
A5.1 The blunder of empiricism
A5.2 The Ten Commandments converted into law
A5.3 A philosophical model of morality
A5.4 Mind-body communication

Appendix 1: God and the world

A1.0 Summary

It shall be shown that, the deeper physicists delve into nature, the more they are convinced of the existence of a single law that rules the totality of matter. Physicists themselves seem to have some trouble drawing the obvious conclusion, but the existence of such a law implies its non-materiality by definition, and begs the question which non-material mind thought of that law. We already convened in calling this mind 'God'. From the physicists' conviction of the existence of a single law ruling the totality of matter (by the way, every other option would be inconsistent) two necessary conclusions must be drawn: that God thought the law, and that He decided to create one material world satisfying that law.

One possible way to get some idea of the existence of a single universal law involves scientific manuals. These are books covering a small but cohesive part of scientific research. The remarkable aspect of these manuals is that they never contradict one another. The only explanation for such a phenomenon is that these manuals must be describing something existent independently from the human mind. The most sublime expression of 'that existent thing' is a mathematical equation describing quantitative law between different properties. Such a mathematical equation exists and is non-material. The materialist responds: "but not everything you think of, or write down, therefore necessarily exists!" True. But the existence is derived from something else: *from the fact that all scientists in the business write down the same equation.* Whenever they forget the equation, they simply look it up in the manual. The materialist responds: "but all those laws you claim exist, are being replaced by more encompassing laws; so how can you claim that Newton's gravity still exists, after Einstein's tank demolished it?" Well, dear materialist, this is exactly how you and I learned to perceive the material part of the world. During the first seconds after having left our mother's womb, we didn't see much of it either, did we? Now, the only difference between the material and spiritual worlds is that the appearance of material part does not change so much anymore when we are adults, while as far as conceiving the spiritual part we are still like a few seconds old. What first looked like Newton's gravity now looks like Einstein's. And the older we get, say one whole minute, the clearer our view on gravity will become: who knows a string-theoretical model will take over. Another comparison is the microscope in biology. That we now see bacteria swimming around, while we did not some centuries earlier, is not an argument against the existence of bacteria. On one condition only: we must be able to understand why without microscope we did not, and with microscope we do see bacteria.

Once that is understood, the simplest model of our experiences is obviously the existence of an objective reality, which mankind learns to appreciate and understand gradually with time.

Throughout this book the terms 'non-material' and 'spiritual' are exchangeable. From a philosophical point of view, the author's view on this world is very similar to that of Aristotle. A more detailed philosophical theory is presented in appendix 4, which tackles the nature of causality. In this appendix the existence of the world is proved on the basis of sciences. In the first section we will first study the history of the manual (A1.1); then its necessary, objective, dogmatic, and universal character (A1.2, A1.3, A1.4, A1.5).

A1.1 A short history of the manual

The seven 'liberal arts' — the trivium (grammar, rhetoric, and logic) and the quadrivium (arithmetic, geometry, music, and astronomy) — were codified in late Antiquity (4th-7th centuries). It was the base for the scholarly curriculum in Western Europe until the translation and subsequent spread of Aristotle's works in the 12th century. Starting from the 18th century, history and natural sciences (physics and chemistry) were gradually introduced. The 19th century witnessed a further decline of the seven liberal arts as compared to natural sciences, and in the 20th century, music, ancient astronomy (the zodiac), logic, and rhetoric were definitively abandoned in favour of the new natural sciences: physics, chemistry, biology and economy.

In the universities the same phenomenon occurred: gradually the emphasis changed from philosophy and theology to the modern sciences.

Today, the annual production of philosophical and theological theses is but a small fraction of those produced in natural sciences. The doctors in science are still called PhD, which is derived from the Latin 'Philosophiae Doctor'. This has little to do with the similarities between the sciences and philosophy: it simply indicates that all natural sciences were born in the 16th -18th centuries, when the majority of the academicians were philosophers.

Today the scholarly curriculum is quite fixed: natural sciences, economy, history, mother language and a foreign language (typically Spanish and English). The university curriculum does not change too much either. The names of the subjects change more than their content, possibly to attract students. This stabilization is explained because the basic notions of the sciences are already known; and the major part (over 99%) of contemporary scientific investigation reduces to the patient and unimpressive application of these basic notions.

All the scientific disciplines know the concept of the 'manual', sometimes called 'bible' among the scientists of a certain discipline. This manual is a work of reference that covers all basic notions of that discipline. The more general a discipline, the more manuals there will be. For example, in quantum mechanics (a very general discipline) there might be a hundred manuals; while in solid-state physics (quantum mechanics applied to solid state, that is to say, to the structure of crystals) there might be about ten; and in some very specialized discipline like infrared fibre laser physics there might be one or two.

A1.2 Consensus or necessity?

What is of utmost importance is that manuals of the same discipline present exactly the same equations. Between them there is not the slightest contradiction. Idealists like Thomas Kuhn like to talk of 'a consensus among the researchers', but the concept of 'consensus' throws the fresh fruits with the rotten ones. There are plenty of examples of consensus in the sciences: for example the metre used in the International System, the English 'inch' (1 inch corresponds exactly to 25.4 millimetres), chemists' Ångstrom (exactly 0.1 nanometres), and physicists' Bohr (a_0 is approximately 0.529 Å). *The consensus refers to the free adherence to any of these units, knowing that this adhesion does not imply a judgement, because the number of possible conventions is infinite.* Depending on the system of units, the equations are written in a different way. *But, once the system of units is specified, there is only one way of writing the equation; and this uniqueness does not proceed from any consensus.* The uniqueness springs from the mathematical essence of material reality. This mathematical essence cannot be either seen or touched: it is strictly non-material.

Kuhn's idealism has its roots in Immanuel Kant, for whom the necessity of physical laws came from the structure of the human mind; furthermore, every man and woman would have the same fundamental mental structures. Kant constructed an impressive edifice on this hypothesis. Previous to the quantum mechanics of Schrödinger and the general relativity of Einstein, Kant reasoned that the mental structures necessarily coincide with the then known physical laws, laws which had very simple graphic interpretation: interaction of contact between voluminous particles in a three dimensional space (height, length, width), with time as a fourth independent dimension.

His theories crashed spectacularly upon the arrival of two new physical discoveries. The first blow was dealt by the Relativity of Einstein (in combination with crucial experiments, like the deviation of light rays in the proximity of heavy bodies, or the 'discovery of the perihelion of Mercury').[15] It demonstrated that space and time are not independent dimensions, but that the relative velocity between two bodies depends on the movement of who is measuring.

The second blow dealt to Kantianism is due to Erwin Schrödinger's quantum mechanics. It permitted John Bell in 1964 to demonstrate that *no material body can cause an event:* all events are caused by some source *exterior to Schrödinger's law,* meaning, outside of material reality. Some scientists speak of the 'collapse of the wave function', but this is already pure philosophical speculation. It has little to do with physics. It is affected by extremely primitive problems like who qualifies as an observer (does a

---

[15] The discovery of the perihelion of Mercury refers to the fact that the orbit of Mercury is not strictly elliptical (like all the orbits of a planet around a very heavy centre, according to Newton's laws), but that it has the shape of a rosette with quasi-elliptic petals. It had already been noticed in the 19th century for the planet Mercury although the effect is so small that it needs 12 million elliptical orbits to complete a rosette. For other planets, including Earth, the advance of the perihelion is even smaller.

cat qualify?) and what happens with the wave function if more observers observe the same particle at the same time. The collapse of the wave function is an ugly philosophical model, which softens the worst pain, but it is the best materialism can do. Appendices A3 and A4 show the beauty of a non-materialistic model.

A1.3 Objectivity and dogma

In the history of sciences, new realities are 'discovered' rather than 'consented to'. There is a reason for the choice of these words. All that is discovered was already there, before it was discovered, while that which is consented to starts to exist at the moment of the consent. Light rays were curving in the vicinity of starts, long before Albert Einstein discovered it. Matter consisted of atoms and molecules, long before Ludwig Boltzmann discovered them. Atoms and molecules had quantified levels of energy, long before Max Planck discovered them. All matter had a wave function associated to it, long before Erwin Schrödinger discovered it. Every quantum material event was caused from outside quantum mechanics, long before John Bell discovered it.

What about convention and consent? They come and they go, scientists do not care. They only care about the mathematical essence of material reality. It is impossible to see it with human eyes; just like an electron, even though nobody doubts its existence today. The apparent 'consent' amongst scientists is a necessary result of a strict selection: students that do not understand how to deduce concrete results of general laws simply do not pass their exams. It is difficult to swallow for scientists that they are as dogmatic as the Catholic Church. The only difference is that the Catholic Church admits being dogmatic, while the scientists —hypocritically— do not.

They rebuke: 'our dogmas are not petrified like those of the Catholic Church, but they change with every scientific revolution'. Well believe me, dear readers, this is nonsense. The key to understanding why this is nonsense is the concept of 'limit'.

- Not a single scientific revolution has contradicted old dogmas: *the old dogmas are always derivable from the new theory in some physical limit;* for example, the old theory of Newton's gravity can be derived from the new one (Einstein's special relativity) in the limit of low velocity (<3000 kilometres per hour).

- In all the history of science a scientific dogma concerning a generally accepted theory has never been falsified; to the contrary, sooner or later it will be derivable from new theory in some specific physical limit.
- Even the dogmas of quantum mechanics, radically distinct from all the dogmas previously known, do not present a single conflict with the dogmas of the previous classical theories; in addition, the old dogmas are perfectly derivable from the quantum ones, in the limit of high temperatures and many particles.

On the other side, it is not true either that catholic dogmas are petrified. Every time that our philosophical understanding of words or concepts changes appreciably, it is necessary to re-formulate the old dogmas; logically, because it is the *content* of the dogma that is of importance to the church, and not its *historical formulation*. The Jews and Christians did not have the luck of the Muslims, whose Allah had the prophet dictated the Koran straight from the sky. The only scripture that came to the Jews directly from the sky are Moses' tablets containing the Ten Commandments. Regrettably, on the first occasion the receiver threw it into pieces. The Lord had mercy, so the Jews received a copy. The efforts of the Nazis and Indiana Jones seem to confirm that the second version did not survive the fall and sack of Jerusalem in the year 587 AD.

A1.4 Philosophical manuals

There also exist manuals for philosophy and theology. With an essential difference: these manuals describe, with certain authority, the opinion of some author(s), but a general consent in terms of the truthfulness of these opinions does not exist. There are always great minds with great authority that differ in opinion from the majority. This phenomenon does not occur in the sciences.

It is well known that Isaac Newton wrote close to a million words about alchemy, spending hours and house of study. The same Newton had his writings about alchemy well stored, probably out of fear from being discovered and ridiculed. The fact is that all of the modern sciences have gradually been substituting dark medieval practices. Astronomy substituted astrology, and modern medicine substituted homeopathic healing. How come philosophy has remained untouched? Philosophers treat each thinker with the same respect. Among all of them *there seems to be not a single charlatan*. We physicists uncover charlatans regularly, the most famous of them being Jan Hendrik Schön.

Other disciplines have their charlatans too: like University of Kentucky biomedical researcher Eric Smart, Japanese anaesthesiologist Yoshitaka Fujii, and forensic chemist Annie Dookhan at the shuttered Department of Public Health Lab in Massachusetts, who was responsible for the incarceration of many innocent persons, the release of many guilty ones, and for millions and millions of public funds spent on rectifying her laziness.

The great mystery is of course who the charlatan philosophers are.

Now let us take the 'father of positivism' Auguste Comte. He conceived three universal historical phases of humanity: theological, metaphysical, and positive. In Comte's language the words 'theology' and 'metaphysics' were but invectives referring to several degrees of retardation, while 'positive' was, uh…, positive. Incidentally, the three 'universal' phases exactly coincide with Comte's century, the 18th: the theological phase corresponds to the despotic monarchism of the Ancien Régime, when 'man blindly believed the existence of an infallible God'; the metaphysical phase corresponds to the revolution (1789), when 'man started to question authority and religion'; and the positive phase corresponds to the post-Napoleonic monarchy, when 'man decided to resolve social problems scientifically'. The only reasonable passages from Comte so much resemble the work of his mentor, Henri de Saint-Simon, that according to the scientific standards of today, it would be considered plagiarism.

Comte even developed a 'positivist religion', with its own calendar, saints, sacraments, priests, and liturgical events. In total Comte's calendar contains 558 names of great men of all times (women probably do not appear). Historic brutes are commemorated, too, for the sake of their 'perpetual execration'. Once in his life, Comte was right: he put Napoleon in the execration category.

Who can take Comte seriously? His good friend, the English utilitarianist John Stuart Mill, distanced himself from 'the later Comte' in unequivocal terms.

A1.5 Science and ideology

In 1927 Einstein's fame was already immense when an unknown Belgian priest dared to propose a radically different opinion concerning the origin of the universe. Notably, it was Einstein, not the priest, who ended up rectifying. This is typical of natural sciences: it is always truth that survives, not fame. This Belgian priest, Georges Lemaître, demonstrated for the first time that Einstein's theory of relativity naturally explains Edwin Hubble's measurements *upon only assuming that the world had an extremely compact beginning in time.*

Hubble measured light from different stars. Their categorization by colour is called 'spectral resolution'. For example, a hot piece of iron emits a continuous spectrum of colours (that is to say, it emits every colour possible). Depending on its temperature it starts with red, then orange, yellow (temperature of the solar surface), green, and even blue (temperature of a welding flame); while a laser, or a LED (light-emitting diode) only emits a few, very narrow lines. These narrow lines are determined by atomic processes: when an electron in an excited orbit relaxes to an inferior orbit it emits radiation of a very specific colour, with an energy equal to the energy difference of the initial and final orbits.

Hubble made two crucial discoveries. In 1923 he discovered that the distant spectra of the solar systems are slightly displaced 'toward the red' (this is technical language for 'toward lower energies'). In 1929 he discovered that spectral lines were more displaced toward the red, the more distant the galaxies. Hubble was not aware of Lemaître's publication in 1927, who explained the origin of such a relation: the Big Bang and its subsequent expansion.

With Einstein, there were also many other heavy-weight critics, like Fred Hoyle. The latter scientist, of great prestige (much more than Lemaître), was generally considered 'father of stationary state': a theory opposed to that of Lemaître for assuming an infinite past of the universe. In the following years, many measurements were published favouring Lemaître's hypothesis. Hoyle continued to defend his monstrous theory until his death in 2001. The final proof against Hoyle's model was that of cosmic radiation observed in 1964 by Penzias and Wilson.

Microwave (also called 'cosmic') radiation from a black body is a prediction of the second generation of models for the Big Bang. According to these models, the primitive universe was a plasm principally composed of mainly protons, neutrons, electrons, and photons. Due to the expansion the temperature of the plasma fell. Some 380,000 years after the Big Bang the plasma's temperature was low enough (3000K) for the electrons (of negative charge) to stably form neutral atoms by joining protons (of positive charge) and neutrons (free of charge). The interaction of radiation (an electromagnetic field) with charged matter is much stronger than that with neutral matter ('Compton dispersion'). While atoms lost their heat quite slowly due to gravity, the average temperature of the photons continued decreasing due to the expansion until reaching an average temperature of 2.7K at the present day (13.7 billion years after the Big Bang). Now this uniform background radiation, typical for a cooling plasma, was exactly what Penzias and Wilson discovered in 1964.

The story of Hoyle typifies the reaction of many cosmologists that were hesitant to confirm a vision of the world that the retrograde Catholic Church had dogmatically proclaimed toward the end of the middle Ages, that is, about four centuries before the birth of the natural sciences. The scientific community owes Hoyle the nickname of 'Big Bang' for Lemaître's theory, at first used pejoratively. Luckily Lemaître's own proposals ('cosmic egg' or 'primeval atom') never made it. Hoyle was not the first one in a long line of ideologists that tried to camouflage the creative implication of the Big Bang. From a philosophic view, it is again interesting to observe that the notion of the Big Bang imposed itself *despite the strong resentments of the great majority of scientists involved.* This is an unequivocal sign of emancipation and credibility of a scientific discipline.

After Hoyle, some scientists proposed cycles of long duration, like those of Aztec and Indian folklore: a possible solution for general relativity would be such that, after a rapid initial expansion, the universe would start to contract until it would destroy itself in a 'big crunch' (great implosion), coinciding with the next Big Bang. Our world would be an infinite repetition of cycles, without a beginning or an end.

Penrose convincingly explained that this model fails due to an argument of entropy (a physical quantity that, in a closed system like our universe, only increases); therefore a 'big crunch' might have the same volume of the Big Bang, but it can impossibly describe the same state. Meanwhile, recent observations have unequivocally demonstrated the accelerated expansion of the universe. This expansion is due to the so called 'black energy', while the contrary tendency (contraction) is due to mass (visible, as well as dark).

A1.6 Certainty and universality of scientific knowledge

Why don't philosophers reach the same level of agreement as the scientists? Quite simply: philosophical truths are even more veiled than the quantitative ones. Once these scientific truths are known, another step of abstraction is needed to reach philosophical truths. As glued to matter the quantitative sciences may seem, the laws they discover are surprisingly general. An impressive example is that, *for disciplines like biochemistry or solid state theory, the fundamental law is (relativistic) quantum mechanics, and more impressively, that this state of affairs will never change.*

One might object: How is it possible to predict something with certainty by means of the natural sciences? Don't new theories spring up daily? Sure. But only very few of them are generally accepted by scientists, mostly because they do not meet the minimum criteria. At this moment we remind the reader of the 'limit' introduced in section A1.3. An anterior (earlier, more primitive) theory can be derived from a posterior (later, more encompassing) theory, and although the posterior theory describes many more phenomena than the anterior theory, within the mathematical derivation limits they yield the same results. Chronologically the anterior theory always comes first, and then the posterior one. Of quantum mechanics the posterior theories are already known by the theorists. They know exactly under which mathematical limits those posterior theories reduce to quantum mechanics. *They also know that all biochemical phenomena satisfy those very limits.* The same applies to solid state theory. This is an example of an absolute, certain, and universal truth that can be derived from quantitative sciences.

Appendix 2: The human spirit

A2.0 Summary

In the first appendix it was argued that the deeper physicists delve into nature, the more they are convinced of the existence of a single universal law ruling all matter at all times. From this it was but a short line to conclude the existence of a God creator of the universe. In this second appendix, the existence of the human spirit is proven; that is to say, that each human person has a spirit which is the source of understanding (A1.1), of intentionality (A1.2), of self-esteem (A1.3), of creativity, synergism, and sociability (A1.4). The proofs are based on fundamental data from mathematics, biology, etymology, and economy, respectively. Summarized in a few paragraphs:

(i) **Mathematics**. In the discipline of number theory, numbers can be related to series of symbols. Most of these series don't make sense, but one may pick out those series that do make sense, and call them "statements". From all statements we can further pick out those that can be derived from the axioms, using the ordinary rules of logic; from the remaining ones we pick out the "formally disprovable" ones, that is, all those whose negation can be derived from the axioms. The truly amazing Gödel theorem claims the existence of a third category of statements. These statements cannot be derived from the axioms, nor can their negation. The claim is universal in the sense that it holds for all possible consistent axiomatic systems: the existence of this third category of statements is a universal property of mathematics and logic, not just some quirk in a very specific branch of number theory. Gödel made things even worse by proving the existence of at least one statement, for every consistent axiomatic system, which belongs to the third category and at the same time is true. Stated in other words: for an axiomatic system to be consistent, there has to be a true statement which cannot be derived from the axioms. Thus, Gödel's law does not discover a hiatus in our knowledge, but it reveals the need of philosophically distinguishing "understanding truth" from "deriving truth". *Understanding truth* is a non-material property exclusive of the human mind. *Deriving truth* does not require any understanding at all; neither of the derived truth, nor of the source truth, nor of the derivation procedure. This is the stupid, mechanical, deterministic jobs that computers are made for.

(ii) **Biology**. Intentionality of the first order is defined as an animal's ability to have intentions. Second order intentionality is an animal's ability to conceive that another animal can have intentions. Third order intentionality is an animal's ability to conceive that another animal dominates second order intentionality. Fourth order intentionality is an animal's ability to conceive that another animal dominates third order intentionality. Etcetera. Zoologists have discovered that the chimpanzee dominates second, but not third order intentionality. However, man dominates all orders of intentionality, without the need of any calculation. If man were a mindless animal (like a computer), evolution would have needed an enormous amount of time (think in billions of years) to program, say, fourth order intentionality. This amount of time was not available, because the definitive ramification between man and the chimpanzee took place 5 to10 million years ago (according to White, 2009). As all animal behaviour is programmed in the fertile egg (matter), man's domination of infinite intentionality cannot come from matter.

(iii) **Anthropology**. From a scientific point of view, it is not known why each individual person claims his or her own rights to be respected by others. Animals don't and they have survived evolution splendidly. To survive, it is more helpful to trample the rights of others than to claim one's own rights. Hence there has never been any genetic pressure in favour of claiming rights throughout our evolutionary history. Without an extremely strong genetic pressure, it is not possible to explain how, out of the 6 billion human beings, all 6 billion would claim rights. Therefore, the claim of fundamental human rights has a spiritual origin.

(iv) **Economy.** It is a verification that leaves someone totally perplexed; while prices of all material products increase as its demand increases, the cost of the labour decreases as demand increases. It is simple to understand why. Six men on our planet can build a nice wooden castle. If they had a seventh available, he could have produced an extra wing to the castle. Six billion men can build enormously high-tech machines. An extra hand would be able to output one extra car per day. This is an unequivocal indication of both the economic creativity of man and of his synergic-social essence: the more men join to carry out a common project, the more the output of a single person rises.

A2.1 Kurt Gödel's Incompleteness Theorem

More than a century ago the eminent German mathematician David Hilbert formulated the last ten interesting mathematical problems left to humanity. The tenth was the so-called 'Entscheidungsproblem' (decision problem): does an algorithm exist which is able to solve all mathematical problems (of a given kind)? He threw a tantrum upon hearing of Gödel's 1930 result [J.W. Dawson, *Logical Dilemmas: The Life and Work of Kurt Gödel*, A.K. Peters 1995]. Gödel's work started a process that would pulverize Hilbert's dream of the complete mathematical domination by man. Hilbert's reaction showed that he immediately understood the implications which so many mathematicians are reluctant to accept: that mathematical truth is a fundamentally superior concept to mathematical deduction. Gödel realized this by demonstrating

(i) that the consistency of an axiomatic system cannot be deduced from its axioms;

(ii) that every consistent axiomatic system always contains a true statement that cannot be derived from the axioms.[16]

In the following we try to give an extremely concise proof of the second statement. Consider a matrix **R** with two entries, the row number

---

[16] Kurt Gödel 1930 *Über formal unentscheidbare Sätze der Principia Mathematica und verwandter Systeme I,* Monatshefte für Mathematk und Physik **38** p172-198

and the column number, both natural numbers. The row orders all possible functions of one argument, and the column orders all possible arguments. According to these definitions, R[10][14] means nothing but the 10th function operating on natural number 14. The incredible feat of 'mathematical thinking' done by Gödel is to associate every matrix number to a number-theoretical statement, For example, R[10][14] is associated with the statement "10>14", and more generally, R[10][m] is associated with the function statement "10>m". R[10][m] is called a "statement function" because it contains the unspecified variable m. Next Gödel defines a set Q containing only those natural numbers k that satisfy the property denoted with a dot (°): R[k][k] cannot be derived from the axioms.

Since the matrix **R** contains all possible statements, there must exist a specific natural number q, which has a specific value that for the moment we don't know (by the way, its exact value would not tell us anything), such that R[q][m] states that m is a member of Q. The existence of q is granted by the fact that **R** contains all possible statements, and this property will be indicated with an asterisk (*). Next Gödel examines the properties of the statement R[q][q], which is known today as the "Gödel number" in mathematical literature:

1. If R[q][q] is false
    then according to (*) q is not member of Q
    then according to (°) R[q][q] can be derived from the axioms
    then R[q][q] is true
2. If R[q][q] is true
    then according to (*) q is member of Q
    then according to (°) R[q][q] cannot be derived from the axioms
3. If R[q][q] can be derived from the axioms
    then R[q][q] is true
    then (cfr. [2] above) R[q][q] cannot be derived from the axioms
4. If the negation of R[q][q] can be derived from the axioms
    then R[q][q] is false
    then (cfr. [1] above) R[q][q] can be derived from the axioms

Taken together, these four logical derivations leave only one possibility for the Gödel number R[q][q]: it is true, but its truth cannot be derived from the axioms.

The fact that from the axioms neither truth nor falsity of a statement can be derived, as is the case in Gödel's number, is called "formal undecidability". Here, "undecidable" refers to the fact that one can establish neither truth nor falsity; and "formal" refers to the rules of derivation and logic within the axiomatic system. The "formally undecidable" status of a proposition should not be confused with a 'third option' in logic, as if logical propositions are either true, false, or undecidable. This would mean the end of logic itself, and a serious misunderstanding of Gödel's work. Propositions are either true or false, and can be either decidable or undecidable. At least three combinations of these two categories exist: decidable and true, decidable and false, undecidable and true. Hence, truth and decidability are independent concepts.

A typical comment of non-experts is that Gödel's way of reasoning is circular. However, Gödel's argument differs notably from typical circular arguments in that it does not lead to inconsistencies. In footnote 15 of Gödel's article proving the incompleteness of number theory, Gödel anticipated on such criticism: "In spite of its appearance, such a proposition is not at all circular, because that proposition asserts in first instance the non-provability of a very specific proposition (namely, the q-th one in an explicit ordering scheme), and only in second instance (in a certain sense casually) it turns out that this specific proposition coincides with itself." This is a very poor translation into English by a Dutch physicist from the original German: "Ein solcher Satz hat entgegen dem Anschein nichts Zirkelhaftes an sich, denn er behauptet zunächst die Unbeweisbarkeit einer ganz bestimmten Formel (nämlich der q-ten in der lexikographischen Anordnung bei einer bestimmten Einsetzung), und erst nachträglich (gewissermaßen zufällig) stellt sich heraus, daß diese Formel gerade die ist, in der er selbst ausgedrückt wurde." The table below illustrates Gödel's point.

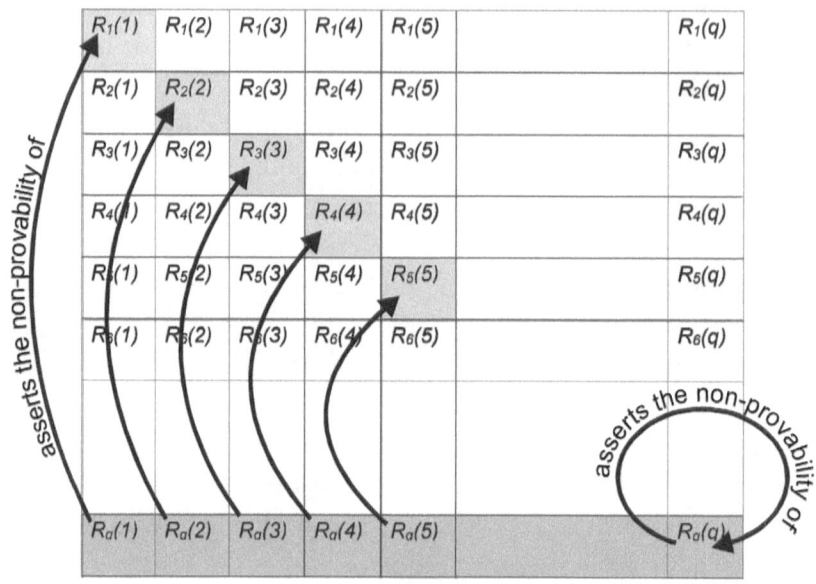

Table: Gödel's 2-dimensional matrix R[k][m], in which every number stands for a specific mathematical statement. They are ordered in such a way that all possible mathematical statements are contained in this matrix. The line number k increases down the vertical axis, and the column number m increases along the horizontal axis. The existence of a specific number q, granted by number theory, defines a whole line of statements (the q-th line), such that R[q] [m] asserts that R[m][m] is not formally provable. This is indicated by means of arrows. The arrow from R[q][3] to R[3][3] indicates that R[q][3] states the formal non-provability of R[3][3], and so forth. Since R[q][m] exists for all values of m, it must also exist for m=q. That is to say, R[q][q] exists, and it truly expresses its own formal non-provability.

The existence of the Gödel number, its truth and formal undecidability, is not some strange property of only Gödel's axiomatic system, but it is a general property of all axiomatic systems. Gödel himself concludes in his original manuscript: "although the proposition is not provable within the axiomatic system, it has been decided by means of meta-mathematical argument". This only makes sense if truth is a concept of a higher level than formal derivability or computation. Computation mechanically links one truth to another, but the process of computation does not require grasping the truth of either premise or conclusion. Formal computation is nothing but a property linking two truths. It follows that computers can never take over human thinking, because the human mind conceives truth, while computing is a mere deterministic process that does not require judgment.

A2.2 Biology: Intentionality

The first modern scientist who suspected that life was nothing but the property of an ingeniously built device was René Descartes [Méditations VI]:

> "And as a clock composed of wheels and weights observes not less exactly all the laws of nature when it is ill-made and does not tell the hours as well as when it is entirely to the wish of the workman, so in like manner I regard the human body as a machine so built and put together of bone, nerve, muscle, vein, blood and skin, that still, although it had no mind, it would not fail to move in all the same ways as at present, since it does not move by the direction of its will, nor consequently by means of the mind, but only by the arrangement of its organs."

Descartes compared the intricate processes undergirding life with the complex machinery needed to keep a clock's minute hand moving around steadily. It took more than three centuries before Descartes' intuition received any significant scientific underpinning. Celebrities like Gregor Mendel and Charles Darwin did the early pioneering work. In 1944 it was shown that the transfer of chromosomes (DNA in its closely packaged form) between different strains of bacteria also transferred heritable properties (B. Alberts, D. Bray, J. Lewis, M. Raff, K. Roberts, and J.D. Watson, *Molecular biology of the cell*, Garland Publishing, New York NY 1995). However, this experimental result was not generally accepted, because scientists did not know of any kind of molecules —apart from proteins—complex enough to carry hereditary information. The proof that those molecules existed and were able to code hereditary information was provided in 1953 by James Watson and Francis Crick. This discovery definitively confirmed the supposition that DNA molecules carry all hereditary information. Nature achieved this admirable feat in a way that is reminiscent of computer programming: though instead of using binary code typical of computers (zeros and ones), nature opted for a four-base code. The four elements making up nature's programming alphabet are the base (in the sense of basic) nucleic acids thymine (T), cytosine (C), adenine (A), and guanine (G). A biological program consists of a long list of these molecules, held together in the form of a strand by a sugar-phosphate backbone. Although many aspects of the function of DNA remain to be elucidated, it is today firmly established that all the hereditary information of an individual is contained in the chromosomes, mainly in the protein-coding parts called genes.

nucleic acid strand (RNA or DNA)

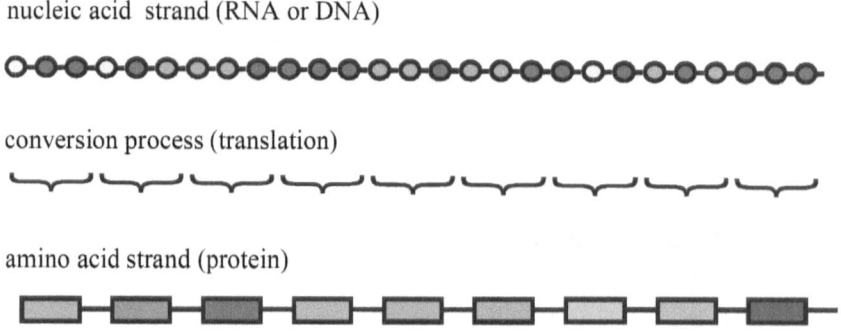

conversion process (translation)

amino acid strand (protein)

Figure A2.2.1: The relation between genes and proteins.

A gene is a strand of nucleic acids coding for a protein. A protein is a strand of amino acids. The four different nucleic acids (A, C, G, and T) are represented by four different grey shades. A triplet of nucleic acids codes for a single amino acid.

Three bases in a row are called a triplet. Since every nucleic acid appears in four possible flavours (A, C, G, T) there exist 4x4x4=64 different triplets. Nature assigned these triplet codes to specific molecules called amino acids, the building blocks of proteins. Since there exist only 20 different amino acids, the 64 possible triplets of nucleic acids imply a large redundancy in the coding. Proteins are chains of amino acids, just like DNA strands are chains of chemically bonded nucleic acids, with the ability to form double stranded helices, made possible by the AC and GT pairing. In this way, nature made it possible to store the structure of all proteins in long strands of just four different basic molecules (see Fig. A2.2.1).

Proteins are extremely useful in biological organisms because they can fold in a very complex, reproducible fashion. Thanks to those complicated three-dimensional structures, proteins can recognise complementary structures in much the same way as a key is 'recognised' by its lock. Proteins are in our own body what assembly robots are in a car factory. The highly specific three-dimensional structure of proteins makes them fit for their role as cellular robots, but it complicates duplication. One could ask at this point: why didn't nature choose to write the program in terms of amino acids, rather than in DNA molecules? Probably the kind of machine able to duplicate all possible proteins would be far too complex. Nature developed an indirect path for protein reproduction. DNA folds as well, but in a standard fashion, which is hardly dependent on the nucleic acids. Thanks to the extremely regular folding pattern of DNA molecules, the metres long DNA strands can condense —with the help of special assembly molecules, like histones— in micron-sized chromosomes. Moreover, the role separation of programming molecules and building blocks allows for a complex regulation of protein expression, which is necessary for complex organisms.

Of course, René Descartes had no inkling of the informational aspects of life, but after all this time it turns out that even the tiniest details of biological life reduce to complex machinery. These are so intelligently designed that it makes one wonder whether blind selection is sufficient to account for their origin (M. Behe, 1996, *Darwin's Black Box*). A nice illustration of protein function is DNA copying. Evidently DNA needs to be copied, and quite often: in every cell division cycle the building plans must be transmitted from mother cell to daughter cell. A unique feature of DNA, discovered shortly before the Watson-Crick model, is its two-strandedness. Two strands of DNA may combine to form a single, two-stranded helix, when all the opposite bases form matched pairs: either A-T or C-G. This feature is exploited by nature to provide for a highly efficient copying process. First, the complementary strands are torn apart, and then special molecular machines —called DNA polymerases— synthesise a complementary strand to both of the original strands. After completion of this process, two identical double-stranded DNA molecules result. This basic process of biological life reduces to the dynamics of charged bodies. Of the four species of nucleic acids only a single one (G) is complementary to C, and likewise for the other three. Whether one molecule can closely approach another is only determined by their spatial charge distributions.

Life, once thought of as some mysterious breath instilled in dead matter, reduces to nothing more than programming intelligence unfolding in a specific hardware environment (like water, salts, lipids, proteins, nucleic acids etc.). Nothing more and nothing less, of course, since the programming of life is extremely complicated: half a century of genetic manipulation has not resulted even once in the improved survival probabilities of a free-ranging organism. All genetic manipulation by man is based on principles already invented by Nature herself, and applied in different circumstances by the scientists.

Though how difficult biological programming may be, it does not invalidate the fundamental parallel between life and mechanism: a biological organism is in all aspects comparable to a self-cloning factory (see Table 3.1). Both need a master plan steering the internal processes, both need complex machinery to perform highly specialised tasks, both need food (prime matter) and throw out excrements (waste), everything in order to yield offspring.

unicellular organism	self-cloning factory
chromosomes	software
proteins	assembly robots
food	raw material
excrements	waste
offspring	produced clones
environment	program input
behaviour	program output
behavioural richness	problem-solving capacity

Table A2.2.2: Basic analogy between a unicellular organism and a self-cloning factory. The last three analogies are explained in the section 'software and genome'.

In spite of all our knowledge concerning the transmittal of hereditary information from parents to offspring, there remains a long way to go before we understand in detail how a given genome gives rise to the corresponding organism. We know how the protein is related to its gene, but we have not yet understood the details of the regulation of gene expression. How a particular cell in the body of some organism looks, depends on what genes are expressed in which cells and with what amounts: the chromosomes of all cells —from the liver, the eye, the skin— of a given organism are roughly identical, but the cells may be completely different. Many years of research are needed before all the details of gene expression are understood in even the simplest organisms. And then there is a second step to make: the transition from the organism's structure to its behaviour. More than a step, this again is a giant leap. Even though we are still very far from understanding the relation between structure and behaviour in detail, it is of fundamental importance to consider that until today there exists not a single piece of evidence indicating that animal behaviour follows from more than molecular structure, and the physical laws governing the behaviour of the molecules. Stated without double negatives: it is perfectly in line with the standards of present-day science to assume that all biological life is fully described by the underlying physical laws. If we define a machine to be any structure whose components are not able, separately, to mimic the behaviour of the whole, then all organisms are machines. Equating organisms with machines may betray a poor sense of romance, but it is without doubt scientifically useful.

The idea that the evolution of the behaviour of organisms is dictated by natural laws, including those strongly anthropomorphic aspects like altruism, care, intimacy, and the like, is one of the major achievements of Darwinism. The road was prepared by Fisher's Genetical Theory of Natural Selection formulated in 1930 (*The Genetical Theory of Natural Selection*, Clarendon Press, Oxford UK) and by Haldane's theory of population genetics dating of 1955 (*Population genetics*, New Biology **18**, 34-51). The definitive proof that a wide spectrum of anthropomorphic behaviour is reducible to natural laws was provided in Hamilton's Genetical Evolution of Social Behaviour in 1964 (W.D. Hamilton 1964 *The genetic evolution of social behaviour*, Journal of Theoretical Biology **7**, p. 1-52). Hamilton explained altruism among insects like Hymenoptera on purely genetic grounds. It is worthwhile to take a close look at Hamilton's argument, explained by Richard Dawkins (*The Selfish Gene*, Oxford University Press, Oxford UK):

> "Insects of the group known as the Hymenoptera, including ants, bees, and wasps, have a very odd system of sex determination. Termites do not belong to this group and they do not share the same peculiarity. A hymenopteran nest typically has only one mature queen. She made one mating flight when young and stored up the sperms for the rest of her long life — ten years or even longer. She rations the sperms out to her eggs over the years, allowing the eggs to be fertilized as they pass out through her tubes. But not all the eggs are fertilized. The unfertilized ones develop into males. A male therefore has no father, and all the cells of his body contain just a single set of chromosomes (all obtained from his mother) instead of a double set (one from the father and one from the mother) as in ourselves. (...)
>
> A female hymenopteran, on the other hand, is normal in that she does have a father, and she has the usual double set of chromosomes in each of her body cells. Whether a female develops into a worker or a queen depends not on her genes but on how she is brought up. That is to say, each female has a complete set of queen-making genes, and a complete set of worker-making genes (or, rather, sets of genes for making each specialized caste of worker, soldier, etc.). Which set of genes is 'turned on' depends on how the female is reared, in particular on the food she receives. (...)
>
> Let us now try to calculate the relatedness between a mother and a son. If a male is known to possess a gene A, what are the chances that his mother shares it? The answer must be 100 per cent, since the male had no father and obtained all his genes from his mother. But now suppose a queen is known to have the gene B. The chance that her son shares the gene is only 50 per cent, since he contains

only half her genes. (…) From a queen's point of view therefore, her offspring, of either sex, are as closely related to her as human children are to their mother.

Things start to get intriguing when we come to sisters. Full sisters not only share the same father: the two sperms that conceived them were identical in every gene. The sisters are therefore equivalent to identical twins as far as their parental genes are concerned. If one female has gene A, she must have got it from either her father or her mother. If she got it from her mother then there is a 50 per cent chance that her sister shares it. But if she got it from her father, the chances are 100 per cent that her sister shares it. Therefore the relatedness between hymenopteran full sisters is not ½ as it would be for normal sexual animals, but ¾.

It follows that a hymenopteran female is more closely related to her full sister than she is to her offspring of either sex. As Hamilton realized (though he did not put it in quite the same way) this might well predispose a female to farm her own mother as an efficient sister-making machine. A gene for vicariously making sisters replicates itself more rapidly than a gene for making offspring directly. Hence worker sterility evolved. It is presumably no accident that true sociality, with worker sterility, seems to have evolved no fewer than eleven times independently in the Hymenoptera and only once in the whole of the rest of the animal kingdom, namely in the termites."

Hamilton was the first to understand that complex social behaviour, like the origin of worker sterility among hymenoptera, is directly related to the genes. Hamilton's reduction can be further reduced to a more fundamental principle, according to which the survival of genes is determined by the ability of self-propagation. This struggle for self-propagation is highly complex because a single gene depends on many others for its propagation. Assuming that genes are the primary objects of evolutionary selection, and that organisms act as their vehicles, it can be understood that the genes of Hymenopteran workers tend to favour their mother's offspring above their own, simply because those workers are more related to their sisters (somewhere between 50% and 75%, depending on the ratio of full sisters to half-sisters) than to their daughters (50%). These numbers explain why worker sterility could originate more than eleven times independently in Hymenoptera, and only once in the rest of the animal kingdom, where the relatedness is equal (twice 50%). Hamilton's final paragraph is very important. It shows why genes are selfish, rather than individuals or species. The reproduction of genes is possible without intentionality — it can be explained with only physical laws. This is not the case for the reproduction of an individual, which cannot be explained on the molecular level. It requires an extra ingredient: intentionality. For example: humans who clone themselves. Apart from this odd phenomenon, evolution in the past 3 billion years did quite well without intentionality.

In 1976 Trivers and Hare argued that, based on Fisher's and Hamilton's theories, the genes of the queen strive for a sex ratio in her offspring of 1:1 while the genes of the female, sterile workers strive for a sex ratio in the queen's offspring of 3:1. These ratios are determined simply by the queen's chances of propagating a given gene to son or daughter (50%:50%=1:1), and by the gene overlap probability between a sterile worker and her sister or brother (75%:25%=3:1). Queen and workers are continually taking measures and counter-measures in order to influence the offspring ratio. Actually, not the queen and the workers, but their genes. Or one step beyond: not the genes, but the natural law dictating the behaviour of genetic molecules.

Having studied twenty species of ants, Trivers and Hare found a close fit to the 3:1 female to male ratio, predicted when the workers win the sex-ratio battle with the queen (Trivers R.L. and H. Hare 1976, *Haplodiploidy and the evolution of the social insects*, Science **191**, 249-263). Looking at two more ant species, of the slave-making kind, they found a sex ratio of 1:1, in accord with the queen's endeavour. Why that? We continue with the words of Dawkins (*The Selfish Gene*, Oxford University Press, Oxford UK, p.178):

> "The consequence of slavery that is interesting from
> our present point of view is this. The queen of the slave-
> making species is now in a position to bend the sex ratio

in the direction she 'prefers'. This is because her own true-born children, the slavers, no longer hold the practical power in the nurseries. This power is now held by the slaves. The slaves 'think' they are looking after their own siblings and they are presumably doing whatever would be appropriate in their own nests to achieve their desired 3:1 bias in favour of sisters. But the queen of the slave-making species is able to get away with counter-measures and there is no selection operating on the slaves to neutralize these counter-measures, since the slaves are totally unrelated to the brood.

For example, suppose that in any ant species, queens 'attempt' to disguise male eggs by making them smell like female ones. Natural selection will normally favour any tendency by workers to 'see through' the disguise. We may picture an evolutionary battle in which queens continually 'change the code', and workers 'break the code'. The war will be won by whoever manages to get more of her genes into the next generation, via the bodies of the reproductives. This will normally be the workers, as we have seen. But when the queen of a slave-making species changes the code, the slave workers cannot evolve any ability to break the code. This is because any gene in a slave worker 'for breaking the code' is not represented in the body of any reproductive individual, and so is not passed on. The reproductives all belong to the slave-making species, and are kin to the queen but not to the slaves. If the genes of the slaves find their way into any reproductives at all, it will be into the reproductives that emerge from the original nest from which they were kidnapped. The slave workers will, if anything, be busy breaking the wrong code! Therefore, queens of a slave-making species can get away with changing their code freely, without there being any danger that genes for breaking the code will be propagated into the next generation."

The successful description of altruism and sex ratios in Hymenoptera represents one of the most spectacular triumphs of what Dawkins later popularised as the 'selfish gene' theory. This theory is extremely sober and elegant, in that it completely dispenses with conscious intentionality. Ants may seem to strive (for certain sex ratios), may seem to sacrifice themselves (in favour of the whole nest), but all such behaviour can perfectly well be explained by natural laws. Given that evolutionary selection does not operate on the level of individuals but of genes, one might ask why individuals care for themselves at all. The answer is simple: only those genes survive, which produce a self-caring behaviour in the individuals. As is repeatedly stressed by Dawkins, such self-caring behaviour is nothing but the blind execution of a genetic program. Self-care does not at all imply the notion of consciousness.

The selfish gene theory is a very successful theory: it reduces a multitude of empirical facts to a limited number of axioms, and its axioms are not inconsistent with the underlying theories (in this case, quantum mechanics and statistical mechanics). The theory may as yet be incomplete —for example, it is not able to explain the stability of species— but additional axioms might account for that in the future.

Another field of research stressing the machine-like origin of altruistic behaviour is game theory. In 1981 Robert Axelrod applied the fundamental principles of game theory to the evolution of co-operation (Axelrod R. 1980 *Effective choice in the prisoner's dilemma*, Journal of Conflict Resolution **24**, p. 379-403), also in situations where kinship does not play a role. In an important paper Axelrod and Hamilton give many examples of how a few elementary rules can give rise to complex strategies (Axelrod R. and W.D. Hamilton 1981 *The evolution of cooperation*, Science **211** p. 1390-1396). The rules are inspired by the 'prisoner's dilemma': if the prisoner co-operates with the plans of his buddy, two things can happen: both might escape from prison, or his buddy might betray him to the guards, in the hope of remission for good conduct. On the other hand, if the prisoner simulates co-operation, but betrays his buddy, the options are the following: either he obtains remission for good conduct, or, if his buddy betrays him too, both get punished. Such a situation can be summed up using only four numbers: the reward for mutual co-operation (both leave the prison), the reward for defection when buddy co-operates (remission for good conduct), the punishment for trying to escape when buddy betrays (sucker's payoff), and the punishment for mutual defection (Table below).

Player A	Player B Cooperation	Defection
Cooperation	P=3 (reward for mutual cooperation)	I=0 (sucker's pay-off)
Defection	T=5 (remission for good conduct)	C=1 (punishment for mutual defection)

Table A2.2.3: The Prisoner's Dilemma game. The payoff to player A is shown with illustrative numerical values.

If Table 2.2.3 were the whole story, every prisoner would always defect, in order not to risk the sucker's payoff. But this is not the whole story. After having defected, the prisoner might once have avoided the sucker's payoff, but then another day in prison starts. Some buddies might be informed by the defection, influencing their behaviour towards our prisoner. This can be quantified by replacing the numbers in Table 2.2.3 by functions with a memory: the prisoner is assumed to be able to remember whether a given prisoner defected the last time, or before last. Depending on the punishment and reward functions, prisoners will adopt different strategies. These strategies can be tested for stability against 'mutants'. That is to say, whenever a stable pattern of behaviour occurs —resulting in time-invariant punishment and reward functions— and some new individuals are introduced in the prison, with totally different behaviour, what might happen to the strategies in use? If these new individuals will either die or adapt their strategies to the prevailing one, then the prevailing strategy is said to be stable.

Such experiments can nicely be simulated on computers. One puts a number of prisoners together, establishes the reward and punishment functions, and follows the total rewards for all prisoners, each one having an own strategy. The Axelrod-Hamilton paper (see their appendix 7 for an extensive quote) argues convincingly that the 'tit for tat' strategy is the best: initial co-operation, and afterwards always repeating one's partner preceding move.

With this basis one can now turn to biological examples, and try to find out what reward and punishment functions can explain the observed strategies. Even though Axelrod and Hamilton do not mention the mathematical form of the reward functions for every biological example, the examples are quite convincing (the reader may judge it by reading the section 'applications' in their appendix 7).

These truly baffling examples of game theory cover a variety of behaviour in many different organisms. They show convincingly how co-operation arises from a limited set of quantitative rules, with no reference to intentionality. Although the surviving strategies are characterised in anthropomorphic terms —forgiving strategies, remorseful, naive, nice, envious— Axelrod and Hamilton stress repeatedly that the players themselves need not be conscious at all.

The level on which games are played in nature naturally depends on the behavioural potentiality of the players involved. Fig trees cannot possibly tell one wasp from another; hence their retaliation must necessarily be 'collective', without distinction between co-operative and defective individuals. However, co-operative bats are able to tell one individual bat from another (Fisher E.A. 1980 *The relationship between mating system and simultaneous hermaphroditism in the coral reef fish Hypoplectrus nigricans Serranidae*, Animal Behaviour **28**, p.620-633). This makes retaliatory strategies much more refined, and the games much richer, much more anthropomorphic.

The behavioural potentiality of the players is determined by their genes, in exactly the same way as the problem-solving capacity of computer programs is determined by the code. This analogy can be condensed in three general programming rules. These can be formulated using the concept behavioural richness. This term signifies all the possible kinds of behaviour obtained when organisms of a given species are allowed to interact with all possible kinds of environments. Due to this definition, there is a univocal relation between a genome and its behavioural richness. A fly does not talk because its genome does not provide it with vocal cords, no matter what kind of environment the fly is exposed to, nor what kind of education it is given.

Behavioural richness of genomes is exactly paralleled by the problem-solving capacity of software programs, and the behaviour of an individual in given circumstances is exactly paralleled by the output of a program for a given input. A software program certainly yields different output when fed with different input —like an organism shows different behaviour when trained or stimulated differently— but its reach of applicability is inherently limited. A tax program yields quite different results depending on whether it is fed with the data of a millionaire or of an unemployed, but it cannot perform reservations for air flights. No matter what the input looks like, the output is either about taxes, or nonsensical.

Without knowing the details of a given program, it is certainly not possible to predict how changes in the source are going to influence the output of the program. Small changes in the software may entail huge changes in the output, and vice versa. Yet it is possible to formulate general rules when we consider only commercial programs, that is, programs that have survived the 'struggle for life' on a healthy, free market.

Three hand-waving software rules for commercial programs:
1. programs differing little in code differ little in problem-solving capacity;
2. programs differing much in problem-solving capacity differ much in code;
3. programs with equal problem-solving capacity may differ much in code.

Subsequent issues of a given commercial program exemplify the first rule. The second rule applies to all possible commercial programs with different applications, like text editing versus numerical integration. The third rule illustrates the first two rules, and is actually redundant. It applies to any couple of source codes with common problem-solving capacity (e.g. compilers), but from different authors. In this case the programs were written independently, and the authors normally use radically different codes. The number of commercial programs available is certainly quite modest for a thorough test of the above three rules, but anybody with a minimum of programming experience knows that new features require new code, and new code is very unlikely the result of slight modifications of whatever old code, written for a different purpose. This principle is at the heart of the three rules.

Analogous rules apply to biological programming as well.

Three hand-waving rules of behavioural richness:

1. genetically similar organisms differ little in behavioural richness;
2. organisms differing much in behavioural richness differ much genetically;
3. organisms with equal behavioural richness may differ much genetically.

This can be verified with the data furnished in the Table A2.2.4. It shows the difference in amino acids of the protein Cytochrome-c, which plays an important role in the energy management in all kinds of organisms, from yeasts and bacteria all the way up to humans. As mentioned in the introduction of this chapter, a triplet of base pairs (in the genome) codes for a single amino acid (of the protein): consequently, mapping the differences in amino acids is an alternative for mapping the genetic differences.

One may look up in the table, for example, that the Cytochrome-c protein of dog and horse differs on 6 of the 104 loci. The first rule (genetically similar organisms differ little in behavioural richness) can be seen to apply for the three birds, for example: penguins, ducks and doves differ at most on 4 loci. The second rule (organisms differing much in behavioural richness differ much genetically) is illustrated by all the white blocks: for example, insects and mammals do not differ on less than 21 loci. The third rule (organisms with equal behavioural richness may differ much genetically) applies to all the grey-background blocks: for example, different types of yeast may differ on as much as 27 loci.

	apes			mammals			birds			fish			insects			plants			yeasts		
	h	c	r	h	d	k	p	d	p	t	b	c	f	s	h	c	s	w	y1	y2	y3
human	0																				
chimp	0	0																			
rhesus	1	1	0																		
horse	12	12	10	0																	
dog	11	11	9	6	0																
kangaroo	10	10	11	7	7	0															
penguin	13	13	12	12	10	10	0														
duck	11	11	10	10	8	10	3	0													
pigeon	12	12	11	11	9	11	4	3	0												
tuna	20	20	20	18	17	17	17	16	17	0											
bonito	20	20	20	17	16	17	17	16	17	2	0										
carp	17	17	17	13	11	13	14	13	14	8	7	0									
fruit fly	27	27	26	22	21	24	24	22	23	23	24	21	0								
silkworm	29	29	28	27	23	26	25	25	25	30	31	25	14	0							
horn worm	29	29	28	26	23	26	25	25	24	28	29	24	13	5	0						
castor	37	37	37	40	38	38	40	38	38	42	41	41	41	40	39	0					
sun flower	38	38	38	41	39	39	41	39	39	43	41	41	41	40	40	10	0				
wheat	38	38	38	41	39	42	41	41	41	44	42	42	42	40	38	12	13	0			
yeast 1	46	46	45	46	45	46	45	45	45	43	42	45	43	43	42	45	47	45	0		
yeast 2	41	41	41	40	38	41	40	40	40	42	41	39	38	39	39	43	44	41	23	0	
yeast 3	41	41	41	42	41	42	40	41	41	43	41	42	42	44	42	42	43	42	25	27	0
bacteria	65	65	64	64	65	66	64	64	64	65	64	64	65	65	64	66	67	66	72	67	69

TABLE A2.2.4: Comparison of the amino acid sequences for Cytochrome-c (104 amino acids). The data are taken from Dayhoff (1972, page D-8). Light grey: small differences. Dark grey: large differences.

The table shows, quantitatively, the difference between how a cytochrome-c molecule is built up from its building blocks (the amino acids) in one organism and another. The diagonal always vanishes, as it compares the species with itself. The grey-tinted squares show the differences within a group of species. For example, the light grey box of apes shows that their cytochromes are much alike, while the dark greys of the yeasts indicate large differences: they have been changing throughout evolutionary history without jumps in species.

All three above rules can be ascertained by the table:
1. penguin, duck, and pigeon differ little in behavioural richness;
2. chimpanzee, fruit fly, and yeast differ much genetically;
3. different yeasts differ much genetically.

Dayhoff's atlas further mentions that humans differ on average from other mammals on 10 loci, from reptiles on 14 loci, from amphibians on 18 loci, from fish on 22 loci. These figures agree qualitatively with fossil data, according to which mammals differentiated 100 million years ago; reptiles, dinosaurs, and birds 300 million years ago; amphibians 350 million years ago; and fish 450 million years ago [Dayhoff 1972, page 48]. Quantitative agreement is quite complicated due to the time and locus dependence of the mutation rates in the genome. The tendency exhibited by Cytochrome-c has been checked for many other proteins, and seems to be universally valid.

In the preceding sections we saw two arguments for the machine-like behaviour of organisms:
- seemingly conscious behaviour like co-operation and self-sacrificing can be understood mechanically, on the basis of the underlying game-theoretical laws, both with and without kinship;
- behavioural richness is related to genome as problem-solving capacity to software.

These insights certainly do not explain all behaviour in nature, but there exists no indication of animal behaviour that is essentially unexplainable in terms of quantitative laws. This is the first ingredient of the proof of incompleteness in biology.

The second ingredient regards the evolution of animal behaviour. In the course of time new species arise and others go extinct. For all species along the human genealogical line (in *latinomics*: eucaryota, metazoa, eumetazoa, bilateria, coelomata, deuterostomia, chordata, craniata, vertebrata, gnathostomata, osteichthes, sarcopterygii, tetrapoda, amniota, synapsida, therapsida, mammalia, eutheria, primates, catarrhini, hominidae, homo), a steady increase in richness of behaviour is readily observable. For every small step forward in behavioural richness, the genetic program (the genome) needs to be slightly adapted. Bigger steps require bigger changes, and bigger changes require longer programming time, just as in software programming. Although nobody knows at present how these changes were effectuated —how new genes came about— from the analogy with software programming it is evident that entirely new behavioural possibilities require completely different genetic systems.

In this section I will briefly review some examples of behavioural richness, with a focus on intentionality. The ethological data are taken from a paper by Sverre Sjölander, published in the Journal of Theoretical Biology in 1997 (*On the evolution of reality: some biological prerequisites and evolutionary stages*, Journal of Theoretical Biology **187** p.595-600), and from the book The Mind of an Ape, by David and Ann Premack (1983 Norton, New York NY).

Stereotyped patterns represent the most primitive way in which nature programs motion of animals. They are characterised by the extremely simple relation between circumstances and behaviour — so simple that the relation may be condensed in a couple of conditionals: if this happens, do this, else do that. In software language: a single if-then-else loop.

Among the many well-documented examples of stereotyped patterns I mention just two:

- Larvae of a dragonfly, Aeschna cyanea, are ambush hunters that camouflage themselves by hiding amidst vegetation. There they wait quietly for a prey to appear (Etienne A.S. 1977 *Descriptive and functional analysis of a stereotyped pattern of locomotion after target tracking in a predatory insect*, Animal behaviour **25** p.429-446). When a larva discovers a prey capable of rapid escape, it pursues the prey with swimming movements, then changes to walking, and finishes with slow creeping movements. If the larva loses visual contact with its would-be prey after having chased it for forty seconds or longer, it performs a three-phased stereotyped pattern: stand still, backward creeping, and pivoting. The function of this pattern is twofold: (i) to stop the larva's unsuccessful attempts to catch a rapidly escaping prey, and (ii) to diminish the probability of future re-encounters with the disappeared prey.

- When a bee has found flowers with plenty of nectar, it takes some of it to the beehive. There the pioneer bee performs the characteristic 'bee dance', consisting of a series of aborted runs toward the flowers. From the details of the dance pattern (dance intensity, dance direction with respect to the sun, run distance, smell of the dancer's body fur) other bees are able to distil information concerning the location of the newly discovered flowers.

The larva's stereotyped pattern evidently does not require the genetic programming of intentionality. However, the bee dance has often been interpreted as a kind of intentional communication. Many animals display the aborted runs of the bee, an example of what is known as 'intention movements'. Birds, for example, perform them by crouching low and spreading their wings, as if they were to fly off. That no real intentionality is involved in the case of the bees is evident from the fact that bees also dance when there are no other bees around to watch it. Yet, many handbooks on animal behaviour still refer to the bee dance as if it were symbolic behaviour in which information is transferred and interpreted intentionally.

Object constancy is acquired by human children of about three years old (J. Piaget 1954 *The Construction of Reality in the Child*, Basic Books, New York, and 1967 *Biologie et Connaissance*, Gallimard, Paris France). Object constancy is the capacity to conceive the identity of an object, even when that object is temporarily hidden from direct eyesight. Such behaviour requires more complicated programming, because it cannot be programmed in a single or a few conditionals (if-then-else loops). Requisites are **memory**, and —for computer software standards— quite advanced three-dimensional object recognition. Etienne's larva mentioned above, for example, is able to anticipate the movement of a prey when it temporarily disappears behind an obstacle (A.S. Etienne, above-quoted article). Such behaviour is very difficult to program for humans, but nobody doubts that Artificial Intelligence will someday succeed in doing so. Intentionality does not seem to be required for primitive forms of object constancy, and is not thinkable without it.

The next step in behavioural richness, on our way towards first order intentionality, is **serial use of senses**, as particularly apparent from the behaviour of reptiles. For example, a snake is able to use different senses in series to catch its prey: that is to say, it is not able to integrate its different senses for a single job. It has to do different parts of the job using the input of a single sense organ at the time. The snake's heat-sensing organs (eyes) govern the striking of the prey. Following the struck prey is governed by smell. The swallowing of the prey is governed by touch. In all three phases, the snake uses the output of only a single sense to determine its behaviour. This is nicely illustrated by the fact that a snake which is hanging from a tree, part of its body coiled around the prey, searches with its head for the prey by bumping into it repeatedly, until it feels the prey's head. Apparently, the snake is not able to use its eyes for locating the prey's head, although its eyes are available, and open. Clearly, the snake's brain can be informed by only a single sense output at the time. Since swallowing is best controlled by touch, all other senses are eclipsed at dinner time.

An evolutionary breakthrough was realised by birds and mammals, which are able, in contrast to reptiles, to integrate different senses at the time. The coordination of senses in a single act is called **intermodality**. From the way in which lions search, hunt, and fight their prey, it is evident that they possess a fantastic capacity of intermodality. Although the associated reality is quite bitter for the prey, filmed sequences are impressive to see, because of the strength and subtlety of athletic performance enabled by intermodality.

First order intentionality is the capacity to have intentions. It is not easy to give a full-fledged definition of this capacity. What one needs at least, is two specific patterns of behaviour (like whining and eating) which are causally connected (a dog whines in order to manifest a basic need) but not 'mechanically' connected: for example, slavering does not manifest intentionality, since it is directly induced by the sight of food in a hungry subject.

Since intermodal animals are intentional too, it is not easy to find out which of the two capacities came first. Possibly, they are two faces of the same coin. For the sake of this chapter's argument, the exact order is of no importance. A nice example of first order intentionality is provided by the behaviour of dogs (S. Sjölander 1995 *Some cognitive breakthroughs in the evolution of cognition and consciousness, and their impact on the biology of language*, Evolution and Cognition **1** p.3-11). A dog that has done something forbidden will meet its master in the door, showing clear signs of distress; or, when thirsty, it will push its water cup around until somebody fills it; and when it's time for the daily walk, it will whine and scratch the door until the master makes some signs of getting ready.

Still, the dog's intentionality is very limited. It is able to anticipate only the immediate future. A dog can communicate its anger at the very moment, but it cannot communicate its feelings of yesterday. Nor can it communicate a conditional, like: 'If you punish me for that, I will chew up the carpet' (Sjölander 1997 *On the evolution of reality: some biological prerequisites and evolutionary stages*, Journal of Theoretical Biology **187** p.595-600). Some clever dogs may be able to deceive, in order to obtain food or escape punishment, though on a primitive level compared to apes. Deception clearly is only possible if the subject conceives that another subject may have intentions. This is the definition of **second order intentionality**. Dogs have it to a certain degree, depending on the breed. Chimpanzees provide convincing examples of second order intentionality (D. Premack and A. Premack 1983 *The Mind of an Ape*, Norton, New York NY). David and Ann Premack showed a chimpanzee a movie of a man jumping for a banana cluster, hanging down from the ceiling. The poor man was not able to reach the bananas. After the movie, the chimpanzee was confronted with some photographs: one with a man standing on a chair, others with selected items shown in the movie, like the banana cluster itself. In ten of the twelve experiments the chimpanzee chose the photograph with the man on the chair. This is quite convincing evidence of the fact that the chimpanzee interprets the man's awkward movements as an attempt to reach for the bananas. It also recognises the chair as a possible solution to the man's problem. Human children younger than four are normally not able to choose intentionally. They choose associatively instead, for the banana cluster, since it also showed up in the movie.

Behaviourists deny the existence of intentionality. They surely must have a hard time explaining plenty of empirical data, like the chimpanzee choices mentioned above. The Premacks are quite unambiguous on the issue of intentionality (D. and A. Premack, page 50):

> "In contradistinction to this widespread popular use of intention, American psychologists (or at least the behaviorists among them) advise us that intention is not only a vague notion, but a bogus one. There is in fact no such thing as intention, the behaviorist assures us; our belief in it is entirely a self-deception. Quite probably, the behaviorists have fallen into the luxury of self-deception on this issue, confusing laboratories with life. In the laboratory, one can easily dispense with intentions. In testing the human subject, instructions can eliminate lying (since human subjects are inclined to tell the truth), so that we do not have to concern ourselves with intentions. And in testing some animals, starvation substitutes for instructions (eliminating any need to speculate about intentions in these experimental situations). After being deprived of food for twenty-four hours (or being maintained at 80 percent of normal body weight), the pigeon or rat whose food is finally restored is likely to do nothing more than eat heartily."

Second order intentionality can give rise to hilarious situations, as in the case of deception among chimpanzees (G. Woodruff and D. Premack 1979, *Internal communication in the chimpanzee: the development of deception*, Cognition 7 p.333-362). Experiments have been carried out whereby one chimpanzee is taught some specific ritual for obtaining food. When the bell rings, the chimpanzee opens a box, fetches a key and opens the container with food. When a second chimpanzee is introduced into the scene, who does not know the trick, a puzzling situation arises for the first chimpanzee. If it simply follows the ritual, it certainly obtains its food, but the second chimpanzee might get to know the trick as well, thereby becoming a potential competitor. So what does it do? When the bell rings, it keeps quiet, as if nothing happened. Only when the second chimpanzee gives up spying, leaving the critical area, will the first chimpanzee approach the box and fetch the key. However, this deception is sometimes counter-deceived: the second chimpanzee feigns walking off, but really hides, peeking intently from behind a tree. When the first chimpanzee opens the box containing the key, it gives away the secret: a deceiving chimpanzee deceived. An example illustrating the limitation of intentionality among free-ranging chimpanzees concerns the intentional act of pointing (D. and A. Premack, page 56):

> "Even though all four young animals developed pointing on a voluntary basis (we did not train them), none has ever used the gesture outside the test space. Even though the animals could have used pointing in numerous situations (to direct buddies to the location of hidden food in the compound, to deceive rivalrous animals, to indicate to their trainers an object or food they wanted), no animal has ever pointed in "free" space. This leads us to conclude that while pointing is not a reflexive behavior, it remains conditioned to the test room; it has not fully undergone the kind of transition that frees it of environmental control. Thus, it has not become a completely spontaneous act that the animal can use, voluntarily, to express its intentions."

Second order intentionality requires some degree of abstraction. I have not found a helpful definition of abstraction in the ethological literature, so the reader has to rely on intuition. Abstraction is involved, for example, in interpreting the meaning of a symbol sequence. Chimpanzees are able to associate certain symbols with subjects and with operations. They might understand a sentence stating 'Sarah give apple Mary', meaning that the research fellow (Sarah) is to hand an apple to the chimpanzee (Mary). Chimpanzees are able to understand the convention that the first-mentioned subject is the donor, and the second one receiver. When Mary has had her fill, she is even able to understand the sentence 'Mary give apple Sarah'. These examples show that chimpanzees are able to abstract, in the sense that they can tell the difference between the object and its symbolical function as a pointer toward other objects (or operations). However, chimpanzees do not grasp the essence of syntax: there is no way to tell them the difference between subject and direct object in a sentence. Without doubt, such a distinction requires a higher level of abstractive power than the one mentioned above, where the operative subject was identified from its leading position in the word sequence.

Two telltale examples illustrating the limitations of chimpanzee abstraction concern map reading and maths. Let me quote the Premacks again (D. and A. Premack, page 103-104):

> "The ape's failure with normal models, such as photos, television, and dollhouses, had forced us to the extreme of using a model that was identical to the world it represented. Finding that the ape could use an identical room as a model helped us to move toward our ultimate objective—the use of a map. With that in mind, we covered the floor of the model room with a canvas sheet, placing the original furniture upon it exactly as it had been. We next reduced the size of the canvas and the size of the furniture. Then we placed the smaller replicas of the original furniture along the perimeter of the canvas (not the edge of the room), for we were trying to use the canvas as a representation of the room next door. Much to our pleasure, these reductions and changes, while at first disruptive, did not result in any permanent changes in the animals' performances. They could use the canvas with its reduced furniture as a guide to the real room. We continued to reduce the size of both canvas and furniture until the furniture became no more than a kind of abstract representation of the originals, arranged on an essentially map-sized piece of canvas. While some disturbances again occurred initially, the apes' performances recovered once more. Could we now use the map outside its original room? And if we changed the orientation of the map, could the ape locate itself appropriately? While two of the animals continued able to use the map as a guide when the map was moved out of the room into the hall (a very minor transfer test), all four animals failed to recognize

new orientations of the map, even when the map was left in the original room and the angle of change was a mere 45 degrees. So, our elaborate training had really accomplished very little. Not one of the animals could be given the piece of canvas in its home cage (with a mark on one of the pieces of furniture) and then travel the few steps down the hall to the test room and find there the concealed food. Even the two animals that could use the map outside the original model room failed abysmally when shown a new map of another, though familiar, space. The training, for whatever its small success, had failed completely to instill the idea of a map."

And with respect to chimpanzee calculus (page 78):

"Sarah was unable to judge whether two rows of buttons were same or different even before we made any changes in their spacing. Sometimes she judged rows equal or same when they were not, then judged rows unequal or different when they were indeed equal. Since she could not judge number in the initial part of the test, how could she judge any transformations we might make in pressing the rows together or in removing an occasional button on tests?"

As one may readily appreciate, the latter experiment is still miles away from calculating sums and differences. Still, Sarah did not qualify. And Sarah is not just a chimpanzee, but no doubt —after rigorous selection and many years of dedicated training— a very privileged one.

Thanks to the chimpanzees' mastering of the second degree of intentionality, and the corresponding abstraction, their behavioural richness enables many more behavioural patterns than other animals. In a synthetic presentation of chimpanzee observation Whiten and coworkers write (A. Whiten, J. Goodall, W.C. McGrew, T. Nishida, V. Reynolds, Y. Sugiyama, C.E.G. Tutin, R.W. Wrangham en C. Boesch, 1999 Nature 399 página 682): "Here we present a systematic synthesis of this information from the seven most long-term studies, which together have accumulated 151 years of chimpanzee observation. This comprehensive analysis reveals patterns of variation that are far more extensive than have previously been documented for any animal species except humans. We find that 39 different behaviour patterns, including tool usage, grooming and courtship behaviours, are customary or habitual in some communities but are absent in others where ecological explanations have been discounted. Among mammalian and avian species, cultural variation has previously been identified only for single behaviour patterns, such as the local dialect of song-birds." These facts are sometimes used to point out that chimpanzees have their own 'culture', in much the same way as also humans have their culture. They normally forget to mention that the chimpanzee populations are genetically much more heterogeneous than humans. This was noted by Woodruff (D.S. Woodruff 1999 *Chimp cultural diversity*, Science **285** p.836):

> "What is often erroneously referred to as "the chimpanzee" comprises at least two well-differentiated allopatric populations that have diverged genetically for more than 1.5 million years. The same heterogeneity is now recognized in "the gorilla" and "the orangutan." There is several times more mitochondrial DNA variation in a single chimpanzee social group than in the entire human species and more sequence variation at chimpanzee nuclear coding (MHC) and noncoding (HOXB6) regions than in humans. *It is perhaps more surprising that there is any cultural variation in our own relatively homogeneous species than that there is any in our far more variable hominoid relatives.*"

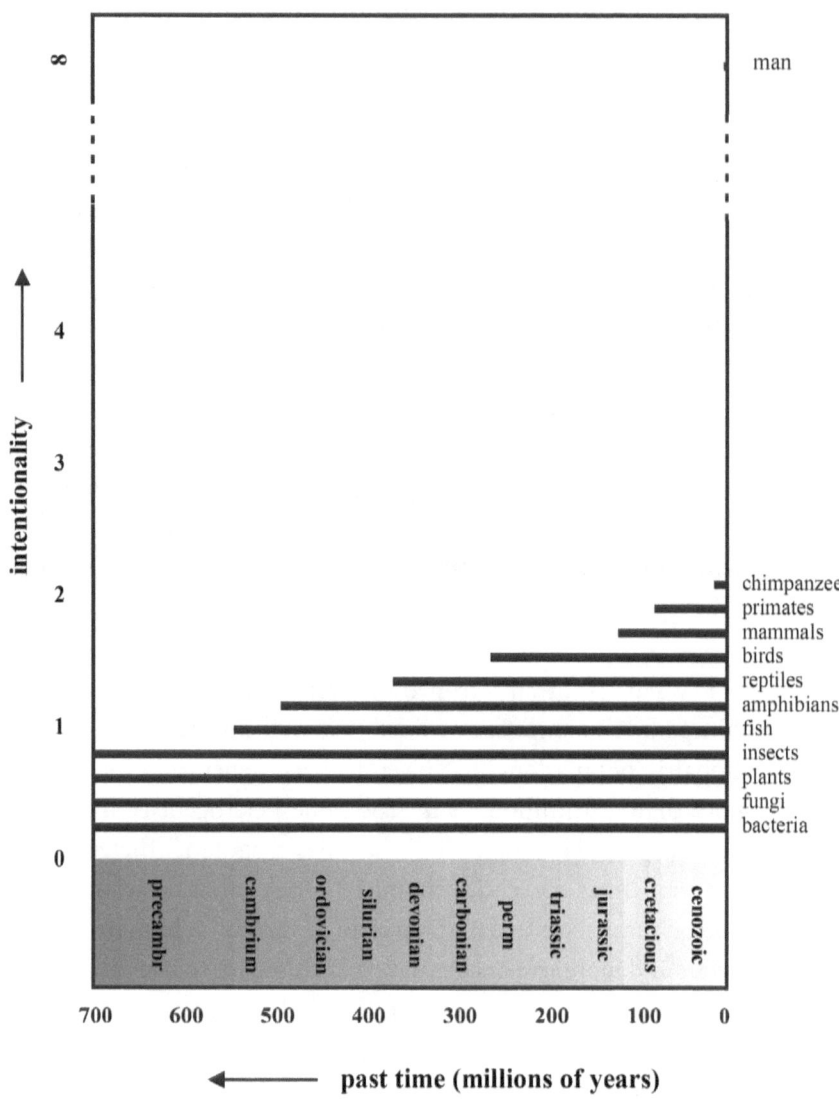

Figure A2.2.5: Order of intentionality as a function of the temporal origin of species.

It is quite impressive that chimpanzees are not able to draw circles, either. Among many thousands of loops drawn by chimpanzees, there was found none with overlapping extremities. These empirical data form the second ingredient of the proof of the incompleteness of biology. Behavioural richness can be quantified using the degree of intentionality. Fig. A2.2.5 plots the animals' degree of intentionality as a function of their moment of birth in history. The subdivision of the first two degrees in intentionality is inspired by the lower stages of behavioural richness: serial use of senses, intermodality, and memory.

The lowest four species (bacteria to insects) were born more than 700 million years ago. The birth of humans (200 thousand years ago) occurred so recently, that the corresponding lifetime bar, drawn to scale, would not be visible in this figure. The interruption of the vertical axes (dashed lines) indicates that human intentionality cannot be accommodated, because it reaches every possible order (symbolised by infinity).

Apparently it took Nature hundreds of millions of years (separating birds from chimpanzees) to program second order intentionality (chimpanzees), once the first order was established (birds). The big interrogative is: how could nature program infinitely many more orders of intentionality in only seven million years — those separating chimpanzees from humans?

We have seen that chimpanzees are capable of full second order intentionality. Maybe they will sometime be found to go even slightly beyond that, but the data make it very unlikely that they will ever reach full third order intentionality. Other primates are known to perform slightly worse than chimpanzees. Non-primate mammals are again slightly worse performing, with dogs located halfway between first and second intentionality. Birds are located close to first order intentionality, and below that all other animals, like the reptiles. But where are we to locate the humans?

Consider a big family, with ten brothers and sisters, in full debate concerning the division of their parents' legacy. Like all his brothers and sisters, John likes to inherit as much as possible. This is John's first order intentionality. John also conceives that all his brothers and sisters want their share —and probably even more than that— of the legacy. This is John's second order intentionality. He also conceives that his sister, Clara, conceives that he himself tries to maximise his inheritance, and that Clara conceives that any other of her sisters and brothers try to do so. This is John's third order intentionality. John also conceives that Clara conceives that Howard conceives that any of his brothers or sisters wants to get hold on the legacy. This is John's fourth order intentionality. A remarkable fact is that any degree of intentionality is accessible for humans. No matter how large the family is, everybody conceives that the first member of the family conceives that the second member conceives that the third member conceives that the fourth member conceives ... that the last member has intentions. An even more remarkable fact is that it does not take a microsecond 'calculation time' longer, for a human, to consider tenth, hundredth, or thousandth order intentionality. In some sense, then, human intentionality should be considered timeless, infinite. The vertical axis of Fig. A2.2.5, however, cannot possibly contain infinity, which is not a natural number. Humans simply do not fit into the figure.

On comparing the genomes of the first animal with second order intentionality (the first chimpanzee) and the first animal with first order intentionality (the first bird, reptile, or amphibian) one may appreciate big genomic differences, and a long time span separating those animals. In terms of the moment of birth, the human and chimpanzee species differ about seven million years. This is less than 5% of the time span separating chimpanzees from birds! Apparently, Nature needed hundreds of millions of years to advance from first to second order intentionality —already an admirable feat— and it should have had enough with only seven million years to program all the other orders of intentionality? If man were a fifth-order intentional animal, his location in Fig. A2.2.5 would already point out a strong discontinuity. The difference is much bigger however: it is infinite...

In spite of the infinite character of the difference between chimpanzee and human behaviour, many people keep looking for structural brain differences. So far, the only identified gene that differs between humans and chimpanzees codes for an enzyme that makes a particular form of sugar called sialic acid: chimpanzees, and all other mammals for that matter, have the gene, while humans do not (J. Alper 2001 *Sugar separates humans from apes*, Science **291** p.2340). Alas for materialism, this brain difference works in the opposite direction. As far as protein expression is concerned, the data are equally deceiving (D. Normile 2001 *Gene expression* *differs in human and chimp brains*, Science **292** p.44-45): the differences are negligible with respect to what they are supposed to explain.

For those who are tempted to think that brain volume is a relevant measure of mental powers: Einstein's brain weighs 2.75 pounds, a quarter pound less than average (E.O. Wilson 1998 *Consilience: The Unity of Knowledge*, A.A. Knopf, New York NY p.97).

These are the data, and they are definitely stunning. There will certainly be no lack of scientists who claim that the riddle is readily solved by assuming that fourth order intentionality is automatically included in third order intentionality, thus providing a necessary continuation throughout all higher orders. What a coincidence that the 'necessary continuation' of intentionality occurs at exactly the third order! Evidently, it would be more realistic to expect such a continuation to occur at second order —an organism able to attribute intentionality to another might well be expected to be able to attribute the attribution of intentionality, too— but the chimpanzee behaviour belies this intuition.

The Premack's do not introduce the multiple orders of intentionality, but rather use just three levels: having intentions, attributing them, and attributing an attribution (A. and D. Premack, p66-67):

> "We have examined intention on two levels: having them and attributing them. There is, in fact, a third level of intention: claiming that others attribute them. That apes have intentions is not questioned; probably most species do. We have just reviewed evidence suggesting that, in addition, chimpanzees may attribute them to us. Can we take the last step and claim the third level, the attribution of attribution? Does Sarah think we attribute intentions to her?
>
> Human children can say of another child who is reaching for a ball that the child wants the ball, or of a child extending her hand in the direction of a lollipop that the child wants the lollipop. But they might have some difficulty when shown the picture of a friend looking at a child reaching for a ball or a lollipop. Would a child predict that his friend would say, "Johnny wants to play

with the ball" or "Johnny wants to eat the lollipop"? The first analysis is of simple attribution; the second is of the attribution of an attribution, which is much more complex. We do not yet know at what age the ability to make such attributions is possible in the child.

Let us describe the kind of test that one must pass in order to claim the third level. In this kind of test, Sarah is no longer shown an actor who has a problem; she is shown an actor observing someone else who has a problem, and she must indicate how the observer will interpret the actor's behavior. In short, the observer in the videotape is now playing the role Sarah formerly played —looking at someone who has a problem— and Sarah is shifted upward to the role of the experimenter.

For example, Sarah is shown Gussie watching Keith jump up and down below a bunch of fruit. The (embedded) videotape that Gussie is watching is stopped or put on hold, and she is shown being offered several photographs of Keith stepping up onto a chair, reaching up with a stick, and so on. Sarah's own videotape is stopped, placed on hold, showing the scene of Gussie confronting the several photographs, and Sarah is presented with several photographs — one showing Gussie selecting a picture in which Keith steps onto a chair, or selecting a picture of Keith reaching out with a stick, and so on. Sarah's task is to choose the photograph depicting the choice that she believes Gussie will make. To perform correctly, Sarah must attribute to Gussie the capacity of attributing intention to Keith, that is, she must attribute an attribution. Human adults can pass this test; but apes cannot. Apes have intentions and probably attribute intentions but the attribution of attribution is restricted to humans."

From this quote it is not clear whether the Premack's identify third order with higher order intentionality, all of them labelled 'attribution of attribution'. I have the impression they do. In my view, such a three-levelled intentionality (having intentions, attributing intentions, and attributing the attribution of intentions) is an erroneous simplification of reality. The only reason to suppose there is no difference between third and higher order intentionality is that we, humans, happen to handle them with equal ease. But the empirical fact is that humans easily handle all finite orders of intentionality, and not the fact that third and fourth orders intentionality are identical. Identifying third and fourth order intentionality —'attribution of attribution' with 'attribution of attribution of attribution'— seems a philosophical error induced by the human extraordinary power of abstraction. In a cheap philosophical context, it is conceivable that the abstraction of some material entity (like a horse) is identical to the abstraction of that abstraction. To put it in another way: the idea of a horse might be identified with the idea of the idea of a horse. Whatever its utility, such a point of view is of no use here: we are not interested in free philosophising, but in finding the model that best explains the empirical data. The above described 'cheap philosophical proposal' to identify higher orders of intentionality simply disregards the empirical fact, well known from computer science, that increasingly complex behaviour requires increasingly complex programming. Fourth order intentionality is a more complex form of behaviour than third order, just like third order is more complex than second and second than first. Whoever claims differently should at least provide a framework within which it can be understood how second and third order intentionality should be programmed. At the time I'm writing these sentences, nobody has been able to program even first order intentionality, and nobody has a clue as to how human or animal brains are programmed to exhibit first order intentionality. How then, can one presume that a program handling third order intentionality necessarily handles fourth and higher orders?

Still, some people claim that the program giving rise to intentionality is simple, and that some 'feedback loop' might provide the miraculous transition from third to arbitrary order intentionality. However, all those aspects of behaviour mentioned above (stereotyped patterns, object constancy, memory, serial use of senses, intermodality) clearly act like prerequisites for intentionality. If not so, one might expect a stone to be fifth-order intentional as well. This is clearly not true, and higher behaviour can only manifest itself when duly based on lower behavioural patterns.

Consequently, even before we have arrived at intentionality at all, Nature already has developed enormously complex software for all the lower behavioural patterns.

Moreover, what would be so special of the feedback loop yielding arbitrary order intentionality? The growth of organisms as 'primitive' as plants is already replete with feedback mechanisms, which operate on the level of DNA copying, on the level of DNA transcription to RNA, on the level of RNA expression into proteins, and on many higher levels that biologists are gradually unveiling. We don't even know on how many levels feedback occurs. How then to explain arbitrary order intentionality by 'just a feedback loop'?

The exceptional character of human intentional behaviour cannot be explained on the basis of natural law — it requires an equally exceptional assumption: a non-material human mind.

A2.3. Anthropology: Subjective claim of fundamental rights

Thomas Aquinas declared that understanding and free will are properties of the mind, because to him it seemed obvious. He never thought of proving the existence of the human mind, nor of its properties. On his time it was not necessary because nobody doubted the existence of the human mind and its properties. Like most people today: lawyers, judges, police, and for many other trades the existence of the human mind is obvious. Not so for physicists, neurologists, and neuropsychiatrists, however. For them the human brain is nothing but a sack of organized molecules. How could a bunch of molecules ever 'contain' the human spirit? Indeed, the brain does not 'contain' the human spirit. One could better consider the brain as the communication channel for the spirit. This is the topic of appendix 4. Now, in order to prove to the nerd (a short for physicists, chemists, neurologists, and neuropsychiatrists) that the human mind exist, there is no other way but starting from their premises: that is to say, we assume that the brain is nothing but a bunch of molecules.

The nerd can verify that animals never discuss moral issues. They can behave as moral beings, as Frans de Waal *claims*, but that is not what he *observes*. He observes nothing more than that chimpanzees display quite complex social patterns. Every computer scientist knows that it is perfectly possible with today's technology to program a bunch of computers in such a way that they display even much more complex social patterns than all chimpanzees, bonobo's, orang-utans, and gorilla's taken together. Clearly, the burden of proving morality in chimpanzees is on De Waal.

But here we can do better. The proof consists of two steps.

First, the chimpanzee and human genomes are so comparable (because they had only 7 million years to diverge) that, if they were identical, it would have been proven that moral judgement does not issue from the brain.

Second, what determines the differentiation of our genomes? Whatever changed in the genome of our common ancestor, who lived some 7 million years ago, must have helped our survival probabilities. Now here comes the gist of the argument. Whatever differentiation the genomes underwent, survivors were selected by their increased adaptation to the natural environment. On one hand, there exists a single moral issue on which all humans agree: that they themselves have fundamental rights. Even Adolf Hitler would subscribe to it. Remember, we are not talking here of human rights, because these refer to the rights of other human beings; here we refer to one's own rights only. Evidently, Adolf Hitler would not subscribe to human rights for Jews. On the other hand, it is well known that humans have very different opinions on moral issues. So how can all humans agree on one single moral issue, when morality issues from the brain, and every brain is different? A theoretical possibility would be that morality is somehow programmed in our brains. This would imply that any change in moral conviction would imply a change in brain structure or brain processes. What then is the reason that no brain change ever occurs due to which a person thinks she has no fundamental rights? Obviously the reason cannot be that one's survival odds depend on one's convictions. All animals known today survived the evolutionary history without ever had any conviction at all. There is no other reasonable option but assuming the existence of human spirit.

A2.4 Economy: Lomborg's Law

The most alarming economic fact in the history of economy is that the price of labour rises with the growth of the density and wealth of the population (cfr. B. Lomborg 2001 *The Skeptical Environmentalist*). This fact can be considered a proof for the existence of the human spirit. Ideologists will simply deny it, just like the phantasmagorical "club of Rome", or its reverend predecessor, Thomas Malthus. In his "Essay on the Principle of Population" Malthus wrote, "When there is no obstacle impeding it, the population doubles every 25 years, growing from period to period, in a geometric progression." He also wrote that, "The livelihoods, in its most prosperous circumstances, will not increase faster than an arithmetical progression."

As exquisite as this may seem, the geometric and arithmetical, it stands in full contradiction to the facts. The growth of a population critically depends on its standard of living and the growth of livelihood critically depends on the level of technology; both of these crucial elements were far beyond Malthus' horizon.

All of the scientific models of population growth make it clear that the global population grows in the shape of an "S", just like the sales volume of each successful product in the free market. According to these models, the world population will saturate at around 12 billion people. The reason for this limit is simple: the more the density of population increases, the more do so the technological level and the standard of living; consequently, there is less need of numerous offspring to survive.

What is the fundamental error of the club of Rome? They consider our planet as a shareable cake — the more we are, the smaller our share. This is a very primitive form of materialism. Thanks to the spirit, the cake grows according to technological developments; today, there is no sign of the limit of the volume of the cake. In rich countries agricultural overproduction is penalized. There is only one exception to this rule: war. Without wars, the earth could easily accommodate a trillion of inhabitants. All together, they would still weigh less than 0.0000000001% of the mass of the earth. With this ratio, for whoever understands the basics of chemistry (an oxygen atom in the brain is identical to an oxygen atom in any piece of trash) it must be evident that the terrestrial population will never be short of food or clothing.

Naturally, there are simpletons that will say that with so many people, there would not be sufficient vital space. This obsession with vital space is typical of racist tyrannies. Let the numbers speak for themselves! The distribution of the density of the world population indicates a strong tendency to concentrate — apparently, Homo sapiens prefers to live in Los Angeles or in Beijing instead of the Mongolian steppe.

Appendix 3: Quantum mechanics

A3.1 Measurement and prediction

Before diving into quantum mechanics it is essential to define our concepts. The standard interpretation of Heisenberg's principle of uncertainty claims that two complementary observables, like position and velocity along a given axis, cannot be measured with arbitrary precision. This standard interpretation is close to the truth, but wrong nonetheless. What Heisenberg's principle claims is that the measurement outcome cannot be predicted with arbitrary precision. The difference between the standard interpretation and the true interpretation is the difference between measurement and prediction. The conclusion is simple: even though the photon's velocity and position along a given axis cannot be predicted with arbitrary precision, they can be measured with arbitrary resolution, and therefore, they are properties that exist at any time. The mistake of the standard interpretation is to assume that only predictable things exist.

For specialists in quantum mechanics a very simple thought experiment illustrates the point.

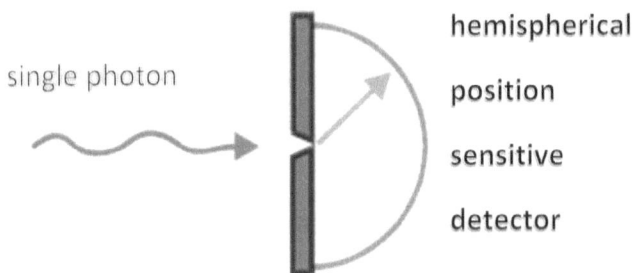

Figure A3.1: The new era of quantum mechanics, or the fundamental loss of full predictive power in physics. A photon in the visible spectrum goes through a small opening (1 micron) to be measured by a spherical detector with good

spatial resolution. The errors of measurement of two complementary variables, like position and velocity of the photon along some axis, are related linearly, and can therefore both be made arbitrarily small.

A3.2 The Copenhagen interpretation

The impossibility of predicting observables is a direct consequence of the wave function: she provides a distribution of possible values to measure, instead of a single specific value. This is completely different from classical mechanics and from Einstein's relativity theories. From what has been said in the section above, it follows that the wave function does not specify the photon's properties, but its *predictable properties*. This interpretation of quantum mechanics differs from the two main variants that were presented at the fifth congress of Solvay in Brussels (1927). On one side, there were the determinists, Einstein and Schrödinger, who were convinced quantum mechanics could be completed by means of a "hidden variable theory". On the other side Werner Heisenberg, Niels Bohr, Wolfgang Pauli, and Paul Dirac believed that quantum mechanics needed no mathematical modification or addition. Subsequent measurements tended to confirm the proposal of Niels Bohr, whence the name "Copenhagen interpretation" (Bohr's university town). Our position differs from the Copenhagen interpretation in the interpretation of the wave function and in the occurrence of its collapse, supposedly as a result of measurement.

A3.3 Correlation

In 1935, Einstein published, together with two co-authors (Podolsky and Rosen), a famous article describing what would later be called the "EPR experiment" of two anti-correlated particles. Consider a source box with a collection of particle pairs, with each pair consisting of a red and a blue particle. When these pairs are measured, "classic anti-correlation" is observed, meaning that if one particle is measured to be red, the other is measured blue; they never have the same colour, simply, because there are no such pairs in the source box. Of course, one can also add a minority of equal-coloured pairs to the source box. In that case the outcomes are called "partially correlated". If there are as many equal-coloured as opposite pairs in the source box, the outcomes are called "totally uncorrelated". For the sake of completeness: if there are only equally-coloured pairs in the source box, the measurements are called "perfectly correlated". These definitions of correlation correspond to common sense.

A3.4 Bell experiments

The awesome, mind-boggling, horrifying, devastating, terrifying truth about quantum mechanics is that quantum sources can be made yielding experimental results under different measurement conditions, whose statistical distributions are incompatible with any classical way of pre-filling the source box.

This possibility was first suggested by a genius called John Stuart Bell. Bell made a relatively simple calculation, with an incredible universality. It took three decades for this miracle to take place. In 1964, Bell published a paper in which he demonstrated that no "local hidden-variable" theory could ever reproduce the quantum predictions.

The calculation had a huge impact. Physicists from all over the world held their breath. Would Bell's calculation herald the end of quantum mechanics, a theory that had allowed so many advances in so short a time?

A frantic hunt on statistics started. After a few years it was clear that quantum mechanics survived, not the human way of thinking about nature. For a long time some die-hards would cling to the possibility of the "detection loophole" (the inconveniences caused by the fact that detectors do not have 100% efficiency), but even this "loophole" was closed, for the first time in 2015 by a group led by Ronald Hanson at Delft University (The Netherlands), using the spins of two entangled electrons over a distance of 1.3 kilometres.

One of the deterministic die-hards is the Dutch Nobel Laureate Gerard 't Hooft. I am not aware of his reaction to Hanson's experiment. As is explained in chapter 5, ideology blinds. In the year 1932, John von Neumann surpassed John Bell, claiming that quantum predictions would never be reproduced by any kind of hidden variable theories. John Bell's claim was more modest, as it referred only to the "local" varieties of those theories, meaning that neither particles nor information were allowed to travel faster than light. In 1952 David Bohm published an explicit theory of non-local hidden variables able to reproduce all possible quantum-mechanical results. John Bell comments, a bit frustrated, "Bohm explicitly demonstrated how to introduce variables into non-relativist quantum mechanics, whose function was to convert the indeterminate description into a determinist one. More important, in my opinion, it was the elimination of subjectivity from the orthodox version, the necessary reference to the "observer". In addition, the essential idea was one already proposed by De Broglie in 1927, with his idea of the 'pilot wave.' (…) Why did Von Neumann not consider it? More extraordinarily, why did people go on producing "impossibility" proofs after 1952, and as recently as 1978? When even Pauli, Rosenfeld, and Heisenberg could produce no more devastating criticism of Bohm's version than to brand it as 'metaphysical' and 'ideological'?"

What exactly does it mean that "local hidden-variable" theories are not able to reproduce the quantum predictions? This means that the mechanism regulating the quantum choices cannot be mimicked by an Einstein-process, in which all particles and signals travel at or below the speed of light. In simpler words: *the quantum choices necessarily come from outside the material universe.* If quantum choices did come from within our universe, issued from whatever processes or objects, then a local hidden variable theory should able to reproduce the quantum predictions. Well, my dear materialists: that is not the case.

So for a materialist the choice is a difficult one: either God, or an ugly and inconsistent wave-function collapse.

Appendix 4: Causality

In this appendix we propose a concept of causality in the context of a more general theory called quantum hylomorphism, which contains elements from both Aristotle's hylomorphism and quantum mechanics.

A4.1 Classical hylomorphism

For Thomas Aquinas an entity in potency is one that does not actually exist, but can exist, while an entity in act is one that already exists. In Aquinas' own words: "Quoniam quoddam potest esse, licet non sit, quoddam vero jam est: illud quod potest esse et non est, dicitur esse potentia; illud autem quod jam est, dicitur esse actu." (Opusculum "De Principiis Naturae ad Fratrem Silvestrum") To be in act, the being in potency has to receive its actuality from outside, "whatever receives something from another is in potency with respect to it; and the received is its act." (St. Thomas, 'De ente et essentia', Ch. 5: "Omne autem quod recipit aliquid ab alio, est in potentia respectu illius; et hoc quod receptum est in eo, est actus eius.")

Act and potency are principles, not things; every singular object, event, or system is constituted by these two principles working together. Act can only multiply entities if they are repeatedly received by potency: potency determines possibilities and, in itself, is not able to produce the transition from possibility to reality. To understand the identity of the philosophical and physical causes, it is of great importance to identify "potency" with the totality of possible states, and "act" with a specific selection by free agent.

In material entities Aristotle considered two compositions: substance-accident and matter-form. Aquinas developed, thanks to the writings of Avicenna and Boethius, a third composition, that of essentia-esse. These components relate to each other as potency to act. At the same time, the essence of a material entity is made up of prime matter and form, which are again related as potency to act (*Quaestio disputata de spiritualibus creaturis*, response to article 1). The unity of being comes from its substantial from, through which it receives being. The teaching on the substantial form is a characteristic of St. Thomas (cfr. John F. Wippel, "The Metaphysical Thought of Thomas Aquinas", 2000 p. 333).

Following Aristotle's footsteps, Aquinas identified prime matter with pure potentiality (the absence of any form or determination). Prime matter can only be known or defined analogically. The Thomist composition of esse and essentia is causal: the actuality of being (esse) is called the transcendental cause, and essence predicamental cause (essentia). The causal dimension of being and essence follows from the Thomist assertion that the concepts of "principle" and "cause" are convertible (cfr. footnote 13, where St. Thomas writes, "Omnis enim causa potest dici principium, et omne principium causa").

Aristotle's hylomorphism was able to philosophically explain the phenomenon of 'change' without loss of identity. Are you the same person after having your hair cut off? Your leg removed? Aristotle's answer would be positive. As we have seen above, according to Aristotle every real thing or phenomenon is philosophically composed of substance and accidents. In case of 'accidental changes' (like the amputation of a leg) the substance remains unchanged, carrying the individuality of the person, while accidents (like the length of the leg) are replaced *with* loss of identity. Aristotle speaks about a 'substantial change' when a living person dies, or when wood is burned to ashes.

A4.2 Quantum hylomorphism

From Aristotle quantum hylomorphism draws the philosophical concept that real things, processes, or phenomena, are composed by two non-material entities which relate like act to potency. From quantum mechanics it draws the necessary evolution of the wave function. Whereas in quantum hylomorphism causality is necessary, in Thomism causality is contingent. "Contingency" is the opposite of "necessity".

Instead of first explaining all the concepts and then showing the quantum hylomorphic composition, we choose to do it the other way around, asking the reader a little bit of patience when concepts are used without having been explained. The tables below give the philosophical composition of man, object or animal, and angel (first column of the composition table). The object is made up of two philosophical 'primary' principles (second column of the composition table). A 'primary' principle is built up of two 'secondary' principles. Following Aristotle, any two principles relate to one another as act to potency. 'Potency' refers to a number of possible ways of being, while 'act' refers to the choice which of these possibilities is realized.

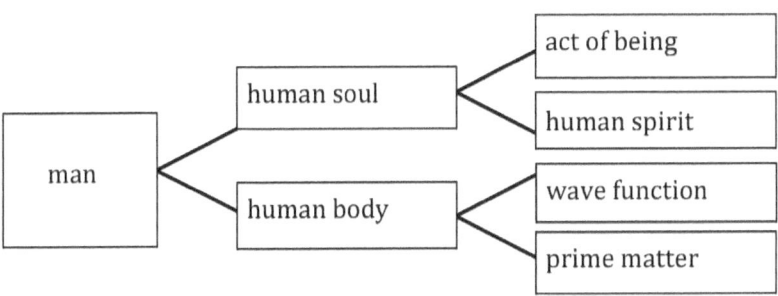

The wave function relates to prime matter as act to potency: prime matter provides 'materiality', but is fully undefined; the wave function specifies, among all possible wave functions, which one makes up a human body. As we speak philosophy here, and not ordinary English, it must be stressed that the human body is not material, but a mere philosophical principle providing 'ordered matter'. Its material properties are identical to that of the man whose composition is shown. The only difference between the two is that 'man' exists in reality and has a spiritual dimension, while the 'human body' is mindless and only exists in the mind of a spiritual being. The spiritual properties of man are due to the human soul. Soul and body relate to one another as act to potency. Soul is made up of the secondary principles 'act of being', which obviously only God can provide, and the human spirit, which again relate to one another as act to potency. The difference between soul and spirit is that the first one exists in reality (although 'incompletely', as it has no body to communicate), while the human spirit does not exist in reality.

From the table it is clear that only the first member of each column has reason of existence. A very technical comment refers to the fact that the human soul is able to determine quantum decisions only in a small part of his brain. All other decisions have to be taken by an angel or by

God: quantum choices in one's feet, legs, liver, stomach, heart, and so forth (fakirs might be an exception to this rule).

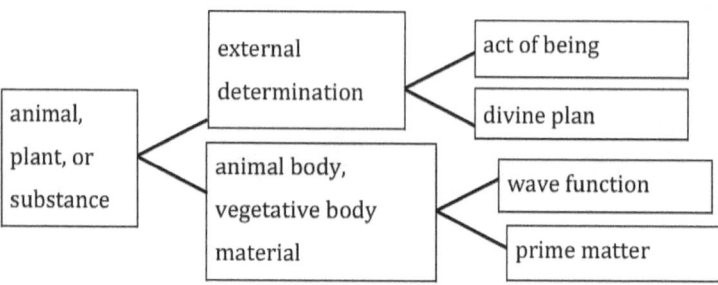

For the rest of material reality, that is, for all non-spirited material reality, the composition table is identical to the human one above, except some of the names. In the third column the human spirit is replaced by the divine plan. Together with the act of being they form the principle of determination, which is responsible for all quantum choices in non-spirited material reality. The potential part of the primary principles of substance (second column, lower branch) can be called 'material organisation' for spirited and non-spirited material beings alike. In all cases, the organisation can be identified with the body, and its information with the wave function. In case of an animal, wave function and soul can be identified. All genetically determined behaviour in humans (animal behaviour) also proceeds from that source: the wave function, or the animal soul of man.

A4.3 Quantum Causality

One does not need to be a genius to appreciate that the second diagram in the hylomorphic context matches one-to-one with the quantum-mechanical world view. Instead of the object one now has the experimental result, and instead of the material organization one now has the experimental set-up.

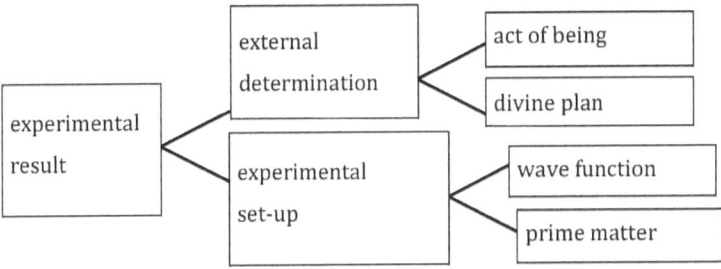

One might even replace the 'experimental result' with 'our actual world', the 'experimental set-up' with 'a possible world', and 'wave function' with 'wave function of the universe'. Indeed, a single wave function suffices to describe the whole universe. It needs never collapse, not even when God Himself looks at it. Its evolution is deterministic because it follows from a Schrödinger-type deterministic equation. This means that the wave function could in principle be known to man.

What is in principle not knowable to man is which of the options of the wave functions will be realized in the future: in appendix A3.4 it was concluded from Bell-type experiments that the source of quantum choices must come from outside material reality. Of course the wave function of the universe contains far too many particles to be of any practical value. What physicists do is devise experiments in such a way that the details of the universal wave function do not matter for the experimental outcome. A typical way to do this was shown in Fig. A3.1: construct a black box with a little hole in it. Whatever the wave function of the universe, the detectors inside the black box will not click unless a photon enters the box through the orifice. Quantum mechanics can then be tested using single-particle wave functions, which are a lot easier to handle than multi-particle wave functions.

There is no collapse of the wave function, neither in the initial nor in the final state. Theoretical physicists 'collapse' the wave function in order to be able to make a calculation. But this collapse is a practical trick which has nothing to do with reality. Those who make the mistake of believing that the wave function collapse is real, end up believing in parallel universes and other concepts of medieval reminiscence.

A4.4 Philosophical causality

The model presented here solves the problem of quantum measurement, and at the time accommodate free will: human, angelic, and divine. From the enormous efforts to domesticate animals one may deduce that the human spirit only has access to a reduced portion of his own brain, just enough to be author of its actions. The angels have access to the remaining animal, vegetable, and inorganic determination. Divinity not only has access to all possible determination, but all also to the creative act of being. Clearly, God plays in a different league.

So what is the crucial effort the present-day philosopher needs to make in order to understand causality in a way that is compatible with the way physics understands causality? The crucial effort is accepting that causes are principles, not beings, and that two of them are needed to produce a single event, and that these two principles relate to one another as act to potency. Honestly, I believe the effort of the physicist to view causality this way requires much more of a 'paradigm change' than for the Thomistic philosopher.

In the following we compare the old and new views on causality, first on the deterministic billiard table, and then in the case of human choice.

The old view of causality on the billiard table is that the movement of one ball causes a change of motion of the other balls. The new view is that the wave function of the billiard table contains but a single outcome. That wave function is one of the two non-material causes. The other one is the external determination, which relates to the wave function as act to potency.

The second example concerns the choice of a man to jump over a ditch, or take a bridge somewhat further away. In this case Thomism would have causality to be "contingent": either thing can happen, without there being a cause favouring one option over the other. In quantum hylomorphism there is no contingency. The two non-material causes are the human body and the human soul, which relate like potency to act. The choice of the human soul is necessarily imprinted in his brain. It depends on the health of the good man if he makes it over the ditch or not.

Appendix 5: Laws of the Spirit

In the second appendix the existence of a non-material human soul is postulated, in order to explain that human behaviour is essentially different from that of a computer. Like the existence of material objects led humanity to acknowledge the existence of quantitative laws governing their behaviour, the existence of spiritual objects (like human souls) led humanity to acknowledge the existence of laws governing the souls' behaviour. It is a deficiency of all European languages that a single word, law, has two separate meanings: in the scientific context it refers to quantitative laws, which have all the characteristics of a Word of God, and in the juridical context it refers to human laws (public, private, criminal, civil, mercantile, constitutional laws etc.). In the chapter we use the term "law" only in the scientific context and "lex" in the juridical context. The differences are:

- law indicates a strict equality between two different quantities;
- lex associates a punishment to offensive behaviour.

They are clearly two distinct concepts, law and lex. Using this terminology it is possible to say that in this appendix the spiritual law is studied, not the spiritual lex. The appendix opens with a discussion of the blunder of empiricism (5.1). It is shown how easily the spiritual law can be derived from divine lex (5.2.). This allows the construction of a model of morality that without naturalistic fallacy (5.3). Then, the communication between matter and spirit will be explained somewhat concisely (5.4). A brief discussion of the nature-nurture duality closes the book (5.5).

A5.1 The blunder of empiricism

Empiricists were not able to conceive the existence of laws. To them every event is a miracle. Of course, the fact that the sun has risen every day during four billion years does not *cause* it to rise tomorrow. However, both events, the sun's rising yesterday and its rising tomorrow, could be due to a cause. In modern language that cause is called the wave function of the universe. In the same way as causes exist on the quantitative level, they also exist on the spiritual level. In the same way that physical law *defines* matter, spiritual law *defines* spirit. Just as the only sensible answer to the question "what is matter" is "all that obeys the universal quantitative law", the only answer to "what is spirit?" is "all that obeys the universal spiritual law." Just like quantitative law can be induced from scientific theory and experiment, spiritual law can be induced from scientific theory and experiment. For example, from the fact that children from broken families run a higher risk to engage in a stable relation, one could postulate a spiritual law that a child's mind benefits from parental love. For believers there exists a much easier way to establish some spiritual laws.

A5.2 The Ten Commandments converted into law

The Ten Commandments can be converted into law by changing the imperative into a consecutive. For example, the fourth commandment commands to honour one's parents. The corresponding spiritual law would be "the more you love your parents, the nobler will be your soul". The law equates two distinct qualities: on one hand, one's free effort spent on loving one's parents, and on the other hand, the quality of one's own soul. The main difference regarding physical law is that spiritual law is not quantitative, but qualitative. The commandment is not law, but lex, because it can be disobeyed, whereas laws rule always and universally, by definition — there is no room for transgression. The laws of the spirit have the same validity and objectivity as the physical laws.

The Russian writer, Fjodor Dostojevski, is a champion of the existence of these spiritual laws. And with him, many classic writers are included, like Miguel de Cervantes Saavedra, Jean-Baptiste Poquelin (Molière), Alessandro Manzoni, Johann Wolfgang von Goethe, William Shakespeare, and Jane Austen. These, and many others, owe their popularity, above all else, to their profound understand of the human behaviour. If this behaviour would be pure randomness, what would be so intriguing about adventures? All that is not pure randomness proves the existence of law.

A5.3 A philosophical model of morality

Once all spiritual laws (hopefully there is but a finite amount) have been established as a result of scientific observations of human behaviour, the important scientific work to be done is to reduce (summarize) all of these laws into a few basic principles, in the same way that in physics, the essential job is to reduce an infinite number of observations into a few principles from which all observations can be derived.

Our proposal contains two fundamental laws:
1. One enriches one's spirit
    (i) the more good one causes for another;
    (ii) the more one forgives personal offenses;

And conversely:
2. The richer one's spirit,
    (i) the happier one will be in this world;
    (ii) the closer one will be to God in the next one.

Clearly, these fundamental laws do not contain Hume's "ought". Fundamental laws do not tell us what we have to do. They only describe how personal action influences the human spirit. Therefore, we have carried out what seemed impossible to Moore: the definition of morality without falling into the naturalistic fallacy.

A5.4 Mind-body communication

The French philosopher, René Descartes, had a dualistic vision of the structure of reality; he proposed that the mind and the body are like to entities of equal philosophical essence that can communicate with each other.

This idea was not compatible with the old physical laws, neither is it with the modern ones (quantum mechanics). As was explained in appendices A4.2 and A4.3, the primary substantial principles (body and soul) of man are both made up of two secondary substantial principles (the act of being and some spirit for the soul, and the wave function and prime matter for the body). This double hylomorphic structure in man

119

allows the expression of a basic idea: on one hand, the wave function and the soul (source of free will) determine which of the allowed quantum option become real. On the other hand, the physical state of (parts of) the brain is "read" by the human soul, source of the five senses: touch, taste, hearing, sight, and smell.

Epilogue

I hereby call upon all righteous people,
believers and non-believers alike,
to unite and fight for peace in our world,
knowing that the enemies are few,
but powerful enough to set the Middle East on fire.

www.ingramcontent.com/pod-product-compliance
Lightning Source LLC
Chambersburg PA
CBHW030808180526
45163CB00003B/1187